The Secretary of the Interior's Standards
for the Treatment of Historic Properties

with Guidelines for Preserving, Rehabilitating, Restoring & Reconstructing Historic Buildings

Kay D. Weeks and Anne E. Grimmer

U.S. Department of the Interior
National Park Service
Cultural Resource Stewardship and Partnerships
Heritage Preservation Services
Washington, D.C.
1995

For sale by the Superintendent of Documents, U.S. Government Printing Office
Internet: bookstore.gpo.gov   Phone: toll free (866) 512-1800;   DC area (202) 512-1800
Fax: (202) 512-2250 Mail: Stop IDCC, Washington, DC 20402-0001

ISBN 0-16-048061-2

Contents

Acknowledgements v

Introduction 1
Choosing an Appropriate Treatment for the Historic Building 1
Using the Standards and Guidelines for a Preservation,
Rehabilitation, Restoration, or Reconstruction Project 2
Historical Overview: Building Materials • Building Features • Site • Setting • Special Requirements 3

Standards for Preservation and Guidelines for Preserving Historic Buildings 17

Introduction 19

Building Exterior: Materials 22
Masonry 22
Wood 26
Architectural Metals 29

Building Exterior: Features 33
Roofs 33
Windows 35
Entrances and Porches 38
Storefronts 40

Building Interior 42
Structural Systems 42
Spaces, Features, and Finishes 44
Mechanical Systems 49

Building Site 51

Setting (District/Neighborhood) 54

Special Requirements 56
Energy Efficiency 56
Accessibility Considerations 58
Health and Safety Considerations 59

Standards for Rehabilitation and Guidelines for Rehabilitating Historic Buildings **61**

Introduction **63**

Building Exterior: Materials **67**
Masonry 67
Wood 71
Architectural Metals 75

Building Exterior: Features **78**
Roofs 78
Windows 81
Entrances and Porches 85
Storefronts 88

Building Interior **91**
Structural Systems 91
Spaces, Features, and Finishes 94
Mechanical Systems 100

Building Site **102**

Setting (District/Neighborhood) **106**

Special Requirements **110**
Energy Efficiency 110
New Additions to Historic Buildings 112
Accessibility Considerations 114
Health and Safety Considerations 115

Standards for Restoration and Guidelines for Restoring Historic Buildings **117**

Introduction **119**

Building Exterior: Materials **122**
Masonry 122
Wood 127
Architectural Metals 131

Building Exterior: Features **135**
Roofs 135
Windows 137
Entrances and Porches 140
Storefronts 143

Building Interior **145**
Structural Systems 145
Spaces, Features, and Finishes. 147
Mechanical Systems 151

Building Site **153**

Setting (District/Neighborhood) **157**

Special Requirements **160**
Energy Efficiency 160
Accessibility Considerations 162
Health and Safety Considerations 163

Standards for Reconstruction and Guidelines for Reconstructing Historic Buildings **165**

Introduction **167**

Research and Documentation **170**

Building Exterior **172**

Building Interior **172**

Building Site **174**

Setting (District/Neighborhood) **175**

Special Requirements **177**
Energy Efficiency 177
Accessibility Considerations 177
Health and Safety Considerations 177

Technical Guidance Publications **179**

Photo Credits

Front and Back Covers

Bangor House, Bangor, Maine, circa 1880. Historic photo (front) and drawing (back): Courtesy, Maine State Historic Preservation Office.

Historical Overview (Materials and Features)

Building Exterior: Masonry. Jack E. Boucher, HABS.

Building Exterior: Wood. Jack E. Boucher, HABS.

Building Exterior: Architectural Metals. Cervin Robinson, HABS.

Building Exterior: Roofs. Jack E. Boucher, HABS.

Building Exterior: Windows. Jack E. Boucher, HABS.

Building Exterior: Entrances and Porches. Jack E. Boucher, HABS.

Building Exterior: Storefronts. Jack E. Boucher, HABS.

Building Interior: Structural Systems. Cervin Robinson, HABS.

Building Interior: Spaces, Features and Finishes. Brooks Photographers, HABS Collection.

Building Interior: Mechanical Systems. National Park Service Files.

Building Site. Jack E. Boucher, HABS.

Setting (District/Neighborhood). Charles Ashton.

Energy Conservation. Laura A. Muckenfuss.

New Additions to Historic Buildings. Rodney Gary.

Accessibility Considerations. Department of Cultural Resources, Raleigh, North Carolina.

Health and Safety Considerations. National Park Service Files.

Chapter Heads

Preservation
Hale House, Los Angeles, California. Photos: Before: National Park Service files; After: Bruce Boehner.

Rehabilitation
Storefront, Painted Post, New York, after rehabilitation. Photo: Kellogg Studio.

Restoration
Camron-Stanford House, Oakland, California. Photos: Before: National Park Service files; After: Courtesy, James B. Spaulding.

Reconstruction
George Washington Memorial House at Washington Birthplace National Monument, Westmoreland County, Virginia. Photo: Richard Frear.

Text

It should be noted that those photographs used to illustrate the guidelines text that are not individually credited in the captions are from National Park Service files.

Revitalization [handwritten]

Acknowledgements

The Standards for the Treatment of Historic Properties, published in 1992, were reviewed by a broad cross-section of government entities and private sector organizations. *The Guidelines for Preserving, Rehabilitating, Restoring and Reconstructing Historic Buildings* were developed in cooperation with the National Conference of State Historic Preservation Officers and reviewed by individual State Historic Preservation Offices nationwide. We wish to thank Stan Graves and Claire Adams, in particular, for their thoughtful evaluation of the new material. Dahlia Hernandez provided administrative support throughout the project.

Finally, this book is dedicated to H. Ward Jandl, whose long-term commitment to historic preservation helped define the profession as we know it today.

The Secretary of the Interior's Standards for the Treatment of Historic Properties may be applied to one historic resource type or a variety of historic resource types; for example, a project may include a complex of buildings such as a house, garage, and barn; the site, with a designed landscape, natural features, and archeological components; structures such as a system of roadways and paths or a bridge; and objects such as fountains and statuary.

Historic Resource Types & Examples

Building: houses, barns, stables, sheds, garages, courthouses, city halls, social halls, commercial buildings, libraries, factories, mills, train depots, hotels, theaters, stationary mobile homes, schools, stores, and churches.

Site: habitation sites, funerary sites, rock shelters, village sites, hunting and fishing sites, ceremonial sites, petroglyphs, rock carvings, ruins, gardens, grounds, battlefields, campsites, sites of treaty signings, trails, areas of land, shipwrecks, cemeteries, designed landscapes, and natural features, such as springs and rock formations, and land areas having cultural significance.

Elmendorf, Lexington, Kentucky. Photo: Charles A. Birnbaum.

Structure: bridges, tunnels, gold dredges, firetowers, canals, turbines, dams, power plants, corn-cribs, silos, roadways, shot towers, windmills, grain elevators, kilns, mounds, cairns, palisade fortifications, earthworks, railroad grades, systems of roadways and paths, boats and ships, railroad locomotives and cars, telescopes, carousels, bandstands, gazebos, and aircraft.

Object: sculpture, monuments, boundary markers, statuary, and fountains.

District: college campuses, central business districts, residential areas, commercial areas, large forts, industrial complexes, civic centers, rural villages, canal systems, collections of habitation and limited activity sites, irrigation systems, large farms, ranches, estates, or plantations, transportation networks, and large landscaped parks.

(Sidebar adapted from National Register Property and Resource Types, p. 15, National Register Bulletin 16A, How to Complete the National Register Form, published by the National Register Branch, Interagency Resources Division, National Park Service, U.S. Department of the Interior, 1991.)

Zoar Historic District, Ohio. Aerial view. Photo: National Park Service.

Introduction

Choosing an Appropriate Treatment for the Historic Building

The Standards are neither technical nor prescriptive, but are intended to promote responsible preservation practices that help protect our Nation's irreplaceable cultural resources. For example, they cannot, in and of themselves, be used to make essential decisions about which features of the historic building should be saved and which can be changed. But once a treatment is selected, the Standards provide philosophical consistency to the work.

Choosing the most appropriate treatment for a building requires careful decision-making about a building's historical significance, as well as taking into account a number of other considerations:

Relative importance in history. Is the building a nationally significant resource—a rare survivor or the work of a master architect or craftsman? Did an important event take place in it? National Historic Landmarks, designated for their "exceptional significance in American history," or many buildings individually listed in the National Register often warrant Preservation or Restoration. Buildings that contribute to the significance of a historic district but are not individually listed in the National Register more frequently undergo Rehabilitation for a compatible new use.

Physical condition. What is the existing condition—or degree of material integrity—of the building prior to work? Has the original form survived largely intact or has it been altered over time? Are the alterations an important part of the building's history? Preservation may be appropriate if distinctive materials, features, and spaces are essentially intact and convey the building's historical significance. If the building requires more extensive repair and replacement, or if alterations or additions are necessary for a new use, then Rehabilitation is probably the most appropriate treatment. These key questions play major roles in determining what treatment is selected.

Proposed use. An essential, practical question to ask is: Will the building be used as it was historically or will it be given a new use? Many historic buildings can be adapted for new uses without seriously damaging their historic character; special-use properties such as grain silos, forts, ice houses, or windmills may be extremely difficult to adapt to new uses without major intervention and a resulting loss of historic character and even integrity.

Mandated code requirements. Regardless of the treatment, code requirements will need to be taken into consideration. But if hastily or poorly designed, a series of code-required actions may jeopardize a building's materials as well as its historic character. Thus, if a building needs to be seismically upgraded, modifications to the historic appearance should be minimal. Abatement of lead paint and asbestos within historic buildings requires particular care if important historic finishes are not to be adversely affected. Finally, alterations and new construction needed to meet accessibility requirements under the Americans with Disabilities Act of 1990 should be designed to minimize material loss and visual change to a historic building.

Using the Standards and Guidelines for a Preservation, Rehabilitation, Restoration, or Reconstruction Project

The Secretary of the Interior's Standards for the Treatment of Historic Properties with Guidelines for Preserving, Rehabilitating, Restoring and Reconstructing Historic Buildings are intended to provide guidance to historic building owners and building managers, preservation consultants, architects, contractors, and project reviewers prior to treatment.

As noted, while the treatment Standards are designed to be applied to all historic resource types included in the National Register of Historic Places—buildings, sites, structures, districts, and objects—the Guidelines apply to *specific* resource types; in this case, buildings.

The Guidelines have been prepared to assist in applying the Standards to all project work; consequently, they are not meant to give case-specific advice or address exceptions or rare instances. Therefore, it is recommended that the advice of qualified historic preservation professionals be obtained early in the planning stage of the project. Such professionals may include architects, architectural historians, historians, historical engineers, archeologists, and others who have experience in working with historic buildings.

The Guidelines pertain to both exterior and interior work on historic buildings of all sizes, materials, and types. Those approaches to work treatments and techniques that are consistent with *The Secretary of the Interior's Standards for the Treatment of Historic Properties* are listed in the **"Recommended"** column on the left; those which are inconsistent with the Standards are listed in the **"Not Recommended"** column on the right.

One chapter of this book is devoted to each of the four treatments: Preservation, Rehabilitation, Restoration, and Reconstruction. Each chapter contains one set of Standards and accompanying Guidelines that are to be used throughout the course of a project. The Standards for the first treatment, *Preservation*, require retention of the greatest amount of historic fabric, along with the building's historic form, features, and detailing as they have evolved over time. The *Rehabilitation* Standards acknowledge the need to alter or add to a historic building to meet continuing or new uses while retaining the building's historic character. The *Restoration* Standards allow for the depiction of a building at a particular time in its history by preserving materials from the period of significance and removing materials from other periods. The *Reconstruction* Standards establish a limited framework for re-creating a vanished or non-surviving building with new materials, primarily for interpretive purposes.

The Guidelines are preceded by a brief historical overview of the primary historic building materials (masonry, wood, and architectural metals) and their diverse uses over time. Next, building features comprised of these materials are discussed, beginning with the exterior, then moving to the interior. Special requirements or work that must be done to meet accessibility requirements, health and safety code requirements, or retrofitting to improve energy efficiency are also addressed here. Although usually not part of the overall process of protecting historic buildings, this work must also be assessed for its potential impact on a historic building.

Historical Overview
Building Exterior Materials

Masonry

Stone is one of the more lasting of masonry building materials and has been used throughout the history of American building construction. The kinds of stone most commonly encountered on historic buildings in the U.S. include various types of sandstone, limestone, marble, granite, slate and fieldstone. *Brick* varied considerably in size and quality. Before 1870, brick clays were pressed into molds and were often unevenly fired. The quality of brick depended on the type of clay available and the brick-making techniques; by the 1870s—with the perfection of an extrusion process—bricks became more uniform and durable. *Terra cotta* is also a kiln-dried clay product popular from the late 19th century until the 1930s. The development of the steel-frame office buildings in the early 20th century contributed to the widespread use of architectural terra cotta. *Adobe*, which consists of sun-dried earthen bricks, was one of the earliest building materials used in the U.S., primarily in the Southwest where it is still popular.

Mortar is used to bond together masonry units. Historic mortar was generally quite soft, consisting primarily of lime and sand with other additives. By the latter part of the 19th century, portland cement was usually added resulting in a more rigid and non-absorbing mortar. Like historic mortar, early *stucco* coatings were also heavily lime-based, increasing in hardness with the addition of portland cement in the late 19th century. *Concrete* has a long history, being variously made of tabby, volcanic ash and, later, of natural hydraulic cements, before the introduction of portland cement in the 1870s. Since then, concrete has also been used in its precast form.

While masonry is among the most durable of historic building materials, it is also very susceptible to damage by improper maintenance or repair techniques and harsh or abrasive cleaning methods.

Wood

Wood has played a central role in American building during every period and in every style. Whether as structural members, exterior cladding, roofing, interior finishes, or decorative features, wood is frequently an essential component of historic buildings.

Because it can be easily shaped by sawing, sanding, planing, carving, and gouging, wood is used for architectural features such as clapboard, cornices, brackets, entablatures, shutters, columns and balustrades. These wooden features, both functional and decorative, are often important in defining the historic character of the building.

Architectural Metals

Architectural metal features—such as cast iron facades, porches, and steps; sheet metal cornices, siding, roofs, roof cresting and storefronts; and cast or rolled metal doors, window sash, entablatures, and hardware—are often highly decorative and may be important in defining the overall character of historic American buildings.

Metals commonly used in historic buildings include lead, tin, zinc, copper, bronze, brass, iron, steel, and to a lesser extent, nickel alloys, stainless steel and aluminum. Historic metal building components were often created by highly skilled, local artisans, and by the late 19th century, many of these components were prefabricated and readily available from catalogs in standardized sizes and designs.

Building Exterior *Features*

Roofs

The roof—with its shape; features such as cresting, dormers, cupolas, and chimneys; and the size, color, and patterning of the roofing material—is an important design element of many historic buildings. In addition, a weathertight roof is essential to the long-term preservation of the entire structure. Historic roofing reflects availability of materials, levels of construction technology, weather, and cost. Throughout the country in all periods of history, *wood shingles* have been used—their size, shape, and detailing differing according to regional craft practices.

European settlers used *clay tile* for roofing at least as early as the mid-17th century. In some cities, such as New York and Boston, clay tiles were popularly used as a precaution against fire. The Spanish influence in the use of clay tiles is found in the southern, southwestern and western states. In the mid-19th century, tile roofs were often replaced by *sheet-metal*, which is lighter and easier to maintain.

Evidence of the use of *slate* for roofing dates from the mid-17th century. Slate has remained popular for its durability, fireproof qualities, and its decorative applications. The use of metals for roofing and roof features dates from the 18th century, and includes the use of *sheet metal, corrugated metal, galvanized metal, tin-plate, copper, lead* and *zinc*.

New roofing materials developed in the early 20th century include built-up roll roofing, and concrete, asbestos, and asphalt shingles.

Windows

Technology and prevailing architectural styles have shaped the history of windows in the United States starting in the 17th century with wooden casement windows with tiny glass panes seated in lead cames. From the transitional single-hung sash in the early 1700s to the true double-hung sash later in the century, these early wooden windows were characterized by small panes, wide muntins, and decorative trim. As the sash thickness increased, muntins took on a thinner appearance as they narrowed in width but increased in thickness.

Changes in technology led to larger panes of glass so that by the mid-19th century, two-over-two lights were common; the manufacture of plate glass in the United States allowed for use of large sheets of glass in commercial and office buildings by the late 19th century. With mass-produced windows, mail order distribution, and changing architectural styles, it was possible to obtain a wide range of window designs and light patterns in sash. Early 20th century designs frequently utilized smaller lights in the upper sash and also casement windows. The desire for fireproof building construction in dense urban areas contributed to the growth of a thriving steel window industry along with a market for hollow metal and metal clad wooden windows.

As one of the few parts of a building serving as both an interior and exterior feature, windows are nearly always an important part of a historic building.

Entrances and Porches

Entrances and porches are quite often the focus of historic buildings, particularly on primary elevations. Together with their functional and decorative features such as doors, steps, balustrades, pilasters, and entablatures, they can be extremely important in defining the overall character of a building. In many cases, porches were energy-saving devices, shading southern and western elevations. Usually entrances and porches were integral components of a historic building's design; for example, porches on Greek Revival houses, with Doric or Ionic columns and pediments, echoed the architectural elements and features of the larger building. Central one-bay porches or arcaded porches are evident in Italianate style buildings of the 1860s. Doors of Renaissance Revival style buildings frequently supported entablatures or pediments. Porches were particularly prominent features of Eastlake and Stick Style houses in which porch posts, railings, and balusters were characterized by a massive and robust quality, with members turned on a lathe. Porches of bungalows of the early 20th century were characterized by tapered porch posts, exposed post and beams, and low pitched roofs with wide overhangs. Art Deco commercial buildings were entered through stylized glass and stainless steel doors.

Storefronts

The earliest extant storefronts in the U.S., dating from the late 18th and early 19th centuries, had bay or oriel windows and provided limited display space. The 19th century witnessed the progressive enlargement of display windows as plate glass became available in increasingly larger units. The use of cast iron columns and lintels at ground floor level permitted structural members to be reduced in size. Recessed entrances provided shelter for sidewalk patrons and further enlarged display areas. In the 1920s and 1930s, aluminum, colored structural glass, stainless steel, glass block, neon, and other new materials were introduced to create Art Deco storefronts.

The storefront is usually the most prominent feature of a historic commercial building, playing a crucial role in a store's advertising and merchandising strategy. Although a storefront normally does not extend beyond the first story, the rest of the building is often related to it visually through a unity of form and detail. Window patterns on the upper floors, cornice elements, and other decorative features should be carefully retained, in addition to the storefront itself.

Building Interior

Structural Systems

The types of structural systems found in the United States include, but are not limited to the following: wooden frame construction (17th c.), balloon frame construction (19th c.), load-bearing masonry construction (18th c.), brick cavity wall construction (19th c.), heavy timber post and beam industrial construction (19th c.), fireproof iron construction (19th c.), heavy masonry and steel construction (19th c.), skeletal steel construction (19th c.), and concrete slab and post construction (20th c.).

If features of the structural system are exposed such as loadbearing brick walls, cast iron columns, roof trusses, posts and beams, vigas, or stone foundation walls, they may be important in defining the building's overall historic character. Unexposed structural features that are not character-defining or an entire structural system may nonetheless be significant in the history of building technology. The structural system should always be examined and evaluated early in the project planning stage to determine its physical condition, its ability to support any proposed changes in use, and its importance to the building's historic character or historical significance.

Spaces, Features, and Finishes

An interior floor plan, the arrangement and sequence of spaces, and built-in features and applied finishes are individually and collectively important in defining the historic character of the building. Interiors are comprised of a series of primary and secondary spaces. This is applicable to all buildings, from courthouses to cathedrals, to cottages and office buildings. Primary spaces, including entrance halls, parlors, or living rooms, assembly rooms and lobbies, are defined not only by their function, but also by their features, finishes, size and proportion.

Secondary spaces are often more functional than decorative, and may include kitchens, bathrooms, mail rooms, utility spaces, secondary hallways, firestairs and office cubicles in a commercial or office space. Extensive changes can often be made in these less important areas without having a detrimental effect on the overall historic character.

Mechanical Systems

Mechanical, lighting and plumbing systems improved significantly with the coming of the Industrial Revolution. The 19th century interest in hygiene, personal comfort, and the reduction of the spread of disease were met with the development of central heating, piped water, piped gas, and network of underground cast iron sewers. Vitreous tiles in kitchens, baths and hospitals could be cleaned easily and regularly. The mass production of cast iron radiators made central heating affordable to many; some radiators were elaborate and included special warming chambers for plates or linens. Ornamental grilles and registers provided decorative covers for functional heaters in public spaces. By the turn of the 20th century, it was common to have all these modern amenities as an integral part of the building.

The greatest impacts of the 20th century on mechanical systems were the use of electricity for interior lighting, forced air ventilation, elevators for tall buildings, exterior lighting and electric heat. The new age of technology brought an increasingly high level of design and decorative art to many of the functional elements of mechanical, electrical and plumbing systems.

The visible decorative features of historic mechanical systems such as grilles, lighting fixtures, and ornamental switchplates may contribute to the overall historic character of the building. Their identification needs to take place, together with an evaluation of their physical condition, early in project planning. On the other hand, mechanical systems need to work efficiently so many older systems, such as compressors and their ductwork, and wiring and pipes often need to be upgraded or entirely replaced in order to meet modern requirements.

Building Site

The building site consists of a historic building or buildings, structures, and associated landscape features within a designed or legally defined parcel of land. A site may be significant in its own right, or because of its association with the historic building or buildings. The relationship between buildings and landscape features on a site should be an integral part of planning for every work project.

Setting (District/Neighborhood)

The setting is the larger area or environment in which a historic property is located. It may be an urban, suburban, or rural neighborhood or a natural landscape in which buildings have been constructed. The relationship of buildings to each other, setbacks, fence patterns, views, driveways and walkways, and street trees together create the character of a district or neighborhood.

Special Requirements

Work that must be done to meet accessibility requirements, health and safety requirements or retrofitting to improve energy efficiency is usually not part of the overall process of protecting historic buildings; rather, this work is assessed for its potential impact on the historic building.

Energy Efficiency

Some features of a historic building or site such as cupolas, shutters, transoms, skylights, sun rooms, porches, and plantings can play an energy-conserving role. Therefore, prior to retrofitting historic buildings to make them more energy efficient, the first step should always be to identify and evaluate existing historic features to assess their inherent energy-conserving potential. If it is determined that retrofitting measures are appropriate, then such work needs to be carried out with particular care to ensure that the building's historic character is retained.

Accessibility Considerations

It is often necessary to make modifications to a historic building so that it will be in compliance with current accessibility code requirements. Accessibility to certain historic structures is required by three specific federal laws: the Architectural Barriers Act of 1968, Section 504 of the Rehabilitation Act of 1973, and the Americans with Disabilities Act of 1990. Federal rules, regulations, and standards have been developed which provide guidance on how to accomplish access to historic areas for people with disabilities. Work must be carefully planned and undertaken so that it does not result in the loss of character-defining spaces, features, and finishes. The goal is to provide the highest level of access with the lowest level of impact.

Health and Safety Considerations

In undertaking work on historic buildings, it is necessary to consider the impact that meeting current health and safety codes (public health, occupational health, life safety, fire safety, electrical, seismic, structural, and building codes) will have on character-defining spaces, features, and finishes. Special coordination with the responsible code officials at the state, county, or municipal level may be required. Securing required building permits and occupancy licenses is best accomplished early in work project planning. It is often necessary to look beyond the "letter" of code requirements to their underlying purpose; most modern codes allow for alterative approaches and reasonable variance to achieve compliance.

Some historic building materials (insulation, lead paint, etc.) contain toxic substances that are potentially hazardous to building occupants. Following careful investigation and analysis, some form of abatement may be required. All workers involved in the encapsulation, repair, or removal of known toxic materials should be adequately trained and should wear proper personal protective gear. Finally, preventive and routine maintenance for historic structures known to contain such materials should also be developed to include proper warnings and precautions.

Standards for Preservation & Guidelines for Preserving Historic Buildings

Preservation is defined as the act or process of applying measures necessary to sustain the existing form, integrity, and materials of an historic property. Work, including preliminary measures to protect and stabilize the property, generally focuses upon the ongoing maintenance and repair of historic materials and features rather than extensive replacement and new construction. New exterior additions are not within the scope of this treatment; however, the limited and sensitive upgrading of mechanical, electrical, and plumbing systems and other code-required work to make properties functional is appropriate within a preservation project.

Standards for Preservation

1. A property will be used as it was historically, or be given a new use that maximizes the retention of distinctive materials, features, spaces, and spatial relationships. Where a treatment and use have not been identified, a property will be protected and, if necessary, stabilized until additional work may be undertaken.

2. The historic character of a property will be retained and preserved. The replacement of intact or repairable historic materials or alteration of features, spaces, and spatial relationships that characterize a property will be avoided.

3. Each property will be recognized as a physical record of its time, place, and use. Work needed to stabilize, consolidate, and conserve existing historic materials and features will be physically and visually compatible, identifiable upon close inspection, and properly documented for future research.

4. Changes to a property that have acquired historic significance in their own right will be retained and preserved.

5. Distinctive materials, features, finishes, and construction techniques or examples of craftsmanship that characterize a property will be preserved.

6. The existing condition of historic features will be evaluated to determine the appropriate level of intervention needed. Where the severity of deterioration requires repair or limited replacement of a distinctive feature, the new material will match the old in composition, design, color, and texture.

7. Chemical or physical treatments, if appropriate, will be undertaken using the gentlest means possible. Treatments that cause damage to historic materials will not be used.

8. Archeological resources will be protected and preserved in place. If such resources must be disturbed, mitigation measures will be undertaken.

Guidelines for Preserving Historic Buildings

Introduction

In **Preservation**, the options for replacement are less extensive than in the treatment, Rehabilitation. This is because it is assumed at the outset that building materials and character-defining features are essentially intact, i.e, that more historic fabric has survived, unchanged over time. The expressed goal of the **Standards for Preservation and Guidelines for Preserving Historic Buildings** is retention of the building's existing form, features and detailing. This may be as simple as basic maintenance of existing materials and features or may involve preparing a historic structure report, undertaking laboratory testing such as paint and mortar analysis, and hiring conservators to perform sensitive work such as reconstituting interior finishes. Protection, maintenance, and repair are emphasized while replacement is minimized.

Identify, Retain, and Preserve Historic Materials and Features

The guidance for the treatment **Preservation** begins with recommendations to identify the form and detailing of those architectural materials and features that are important in defining the building's historic character and which must be retained in order to preserve that character. Therefore, guidance on *identifying, retaining, and preserving* character-defining features is always given first. The character of a historic building may be defined by the form and detailing of exterior materials, such as masonry, wood, and metal; exterior features, such as roofs, porches, and windows; interior materials, such as plaster and paint; and interior features, such as moldings and stairways, room configuration and spatial relationships, as well as structural and mechanical systems; and the building's site and setting.

Stabilize Deteriorated Historic Materials and Features as a Preliminary Measure

Deteriorated portions of a historic building may need to be protected thorough preliminary stabilization measures until additional work can be undertaken. *Stabilizing* may include structural reinforcement, weatherization, or correcting unsafe conditions. Temporary stabilization should always be carried out in such a manner that it detracts as little as possible from the historic building's appearance. Although it may not be necessary in every preservation project, stabilization is nonetheless an integral part of the treatment **Preservation**; it is equally applicable, if circumstances warrant, for the other treatments.

Protect and Maintain Historic Materials and Features

After identifying those materials and features that are important and must be retained in the process of **Preservation** work, then *protecting and maintaining* them are addressed. Protection generally involves the least degree of intervention and is preparatory to other work. For example, protection includes the maintenance of historic materials through treatments such as rust removal, caulking, limited paint removal, and re-application of protective coatings; the cyclical cleaning of roof gutter systems; or installation of fencing, alarm systems and other temporary protective measures. Although a historic building will usually require more extensive work, an overall evaluation of its physical condition should always begin at this level.

Repair (Stabilize, Consolidate, and Conserve) Historic Materials and Features

Next, when the physical condition of character-defining materials and features requires additional work, *repairing* by *stabilizing, consolidating, and*

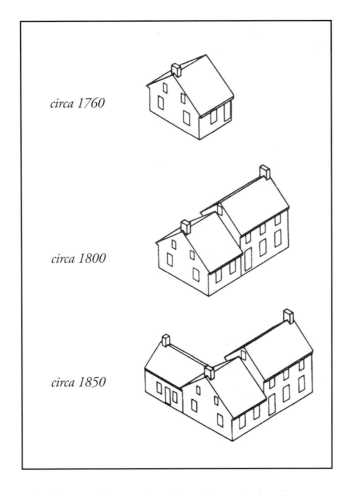

This three-part drawing shows the evolution of a farm house over time. Such change is part of the history of the place and is respected within the treatment, Preservation. Drawing: Center for Historic Architecture and Engineering, University of Delaware (adapted from Preservation Brief 35: Understanding Old Buildings).

conserving is recommended. **Preservation** strives to retain existing materials and features while employing as little new material as possible. Consequently, guidance for repairing a historic material, such as masonry, again begins with the least degree of intervention possible such as strengthening fragile materials through consolidation, when appropriate, and repointing with mortar of an appropriate strength. Repairing masonry as well as wood and architectural metal features may also include patching, splicing, or otherwise reinforcing them using recognized preservation methods. Similarly, within the treatment **Preservation**, portions of a historic structural system could be reinforced using contemporary materials such as steel rods. All work should be physically and visually compatible, identifiable upon close inspection and documented for future research.

Limited Replacement In Kind of Extensively Deteriorated Portions of Historic Features

If repair by stabilization, consolidation, and conservation proves inadequate, the next level of intervention involves the *limited replacement in kind* of extensively deteriorated or missing parts of features when there are surviving prototypes (for example, brackets, dentils, steps, plaster, or portions of slate or tile roofing). The replacement material needs to match the old both physically and visually, i.e., wood with wood, etc. Thus, with the exception of hidden structural reinforcement and new mechanical system components, substitute materials are not appropriate in the treatment **Preservation**. Again, it is important that all new material be identified and properly documented for future research.

If prominent features are missing, such as an interior staircase, exterior cornice, or a roof dormer, then a Rehabilitation or Restoration treatment may be more appropriate.

Energy Efficiency/Accessibility Considerations/Health and Safety Code Considerations

These sections of the **Preservation** guidance address work done to meet accessibility requirements and health and safety code requirements; or limited retrofitting measures to improve energy efficiency. Although this work is quite often an important aspect of preservation projects, it is usually not part of the overall process of protecting, stabilizing, conserving, or repairing character-defining features; rather, such work is assessed for its potential negative impact on the building's character. For this reason, particular care must be taken not to obscure, damage, or destroy character-defining materials or features in the process of undertaking work to meet code and energy requirements.

***Preservation as a Treatment.** When the property's distinctive materials, features, and spaces are essentially intact and thus convey the historic significance without extensive repair or replacement; when depiction at a particular period of time is not appropriate; and when a continuing or new use does not require additions or extensive alterations, Preservation may be considered as a treatment. Prior to undertaking work, a documentation plan for Preservation should be developed.*

Preservation

Building Exterior

Masonry: Brick, stone, terra cotta, concrete, adobe, stucco, and mortar

Recommended

Identifying, retaining, and preserving masonry features that are important in defining the overall historic character of the building such as walls, brackets, railings, cornices, window architraves, door pediments, steps, and columns; and details such as tooling and bonding patterns, coatings, and color.

Stabilizing deteriorated or damaged masonry as a preliminary measure, when necessary, prior to undertaking appropriate preservation work.

Protecting and maintaining masonry by providing proper drainage so that water does not stand on flat, horizontal surfaces or accumulate in curved decorative features.

Cleaning masonry only when necessary to halt deterioration or remove heavy soiling.

Carrying out masonry surface cleaning tests after it has been determined that such cleaning is appropriate. Tests should be observed over a sufficient period of time so that both the immediate and the long range effects are known to enable selection of the gentlest method possible.

Not Recommended

Altering masonry features which are important in defining the overall historic character of the building so that, as a result, the character is diminished.

Replacing historic masonry features instead of repairing or replacing only the deteriorated masonry.

Applying paint or other coatings such as stucco to masonry that has been historically unpainted or uncoated.

Removing paint from historically painted masonry.

Changing the type of paint or coating or its color.

Failing to stabilize deteriorated or damaged masonry until additional work is undertaken, thus allowing further damage to occur to the historic building.

Failing to evaluate and treat the various causes of mortar joint deterioration such as leaking roofs or gutters, differential settlement of the building, capillary action, or extreme weather exposure.

Cleaning masonry surfaces when they are not heavily soiled, thus needlessly introducing chemicals or moisture into historic materials.

Cleaning masonry surfaces without testing or without sufficient time for the testing results to be of value.

Preservation

Recommended

Cleaning masonry surfaces with the gentlest method possible, such as low pressure water and detergents, using natural bristle brushes.

Inspecting painted masonry surfaces to determine whether repainting is necessary.

Removing damaged or deteriorated paint only to the next sound layer using the gentlest method possible (e.g., hand-scraping) prior to repainting.

Applying compatible paint coating systems following proper surface preparation.

Repainting with colors that are historically appropriate to the building and district.

Evaluating the existing condition of the masonry to determine whether more than protection and maintenance are required, that is, if repairs to masonry features will be necessary.

Repairing, stabilizing, and conserving fragile masonry by using well-tested consolidants, when appropriate. Repairs should be physically and visually compatible and identifiable upon close inspection for future research.

Not Recommended

Sandblasting brick or stone surfaces using dry or wet grit or other abrasives. These methods of cleaning permanently erode the surface of the material and accelerate deterioration.

Using a cleaning method that involves water or liquid chemical solutions when there is any possibility of freezing temperatures.

Cleaning with chemical products that will damage masonry, such as using acid on limestone or marble, or leaving chemicals on masonry surfaces.

Applying high pressure water cleaning methods that will damage historic masonry and the mortar joints.

Removing paint that is firmly adhering to, and thus protecting, masonry surfaces.

Using methods of removing paint which are destructive to masonry, such as sandblasting, application of caustic solutions, or high pressure waterblasting.

Failing to follow manufacturers' product and application instructions when repainting masonry.

Using new paint colors that are inappropriate to the historic building and district.

Failing to undertake adequate measures to assure the protection of masonry features.

Removing masonry that could be stabilized, repaired and conserved; or using untested consolidants and untrained personnel, thus causing further damage to fragile materials.

Preservation

Adequate protection and maintenance of a historic building is an ongoing commitment. Here, two workers are priming and repainting exterior stone and wood trim. If surface treatments are neglected, more extensive repair and replacement will be required. Each loss further undermines a building's historic integrity.

Recommended

Repairing masonry walls and other masonry features by repointing the mortar joints where there is evidence of deterioration such as disintegrating mortar, cracks in mortar joints, loose bricks, damp walls, or damaged plasterwork.

Removing deteriorated mortar by carefully hand-raking the joints to avoid damaging the masonry.

Duplicating old mortar in strength, composition, color, and texture.

Duplicating old mortar joints in width and in joint profile.

Not Recommended

Removing nondeteriorated mortar from sound joints, then repointing the entire building to achieve a uniform appearance.

Using electric saws and hammers rather than hand tools to remove deteriorated mortar from joints prior to repointing.

Repointing with mortar of high portland cement content (unless it is the content of the historic mortar). This can often create a bond that is stronger than the historic material and can cause damage as a result of the differing coefficient of expansion and the differing porosity of the material and the mortar.

Repointing with a synthetic caulking compound.

Using a "scrub" coating technique to repoint instead of traditional repointing methods.

Changing the width or joint profile when repointing.

Preservation

Recommended

Repairing stucco by removing the damaged material and patching with new stucco that duplicates the old in strength, composition, color, and texture.

Using mud plaster as a surface coating over unfired, unstabilized adobe because the mud plaster will bond to the adobe.

Cutting damaged concrete back to remove the source of deterioration (often corrosion on metal reinforcement bars). The new patch must be applied carefully so it will bond satisfactorily with, and match, the historic concrete.

Repairing masonry features by patching, piecing-in, or otherwise reinforcing the masonry using recognized preservation methods. The new work should be unobtrusively dated to guide future research and treatment.

Applying new or non-historic surface treatments such as water-repellent coatings to masonry only after repointing and only if masonry repairs have failed to arrest water penetration problems.

Not Recommended

Removing sound stucco; or repairing with new stucco that is stronger than the historic material or does not convey the same visual appearance.

Applying cement stucco to unfired, unstabilized adobe. Because the cement stucco will not bond properly, moisture can become entrapped between materials, resulting in accelerated deterioration of the adobe.

Patching concrete without removing the source of deterioration.

Removing masonry that could be repaired, using improper repair techniques, or failing to document the new work.

Applying waterproof, water repellent, or non-historic coatings such as stucco to masonry as a substitute for repointing and masonry repairs. Coatings are frequently unnecessary, expensive, and may change the appearance of historic masonry as well as accelerate its deterioration.

The following work is highlighted to indicate that it represents the greatest degree of intervention generally recommended within the treatment **Preservation***, and should only be considered after protection, stabilization, and repair concerns have been addressed.*

Recommended

Limited Replacement in Kind

Replacing in kind extensively deteriorated or missing parts of masonry features when there are surviving prototypes such as terra-cotta brackets or stone balusters. The new work should match the old in material, design, color, and texture; and be unobtrusively dated to guide future research and treatment.

Not Recommended

Replacing an entire masonry feature such as a column or stairway when limited replacement of deteriorated and missing parts is appropriate.

Using replacement material that does not match the historic masonry feature; or failing to properly document the new work.

Preservation

Building Exterior

Wood: Clapboard, weatherboard, shingles, and other wooden siding and decorative elements

Recommended ✓

Identifying, retaining, and preserving wood features that are important in defining the overall historic character of the building such as siding, cornices, brackets, window architraves, and doorway pediments; and their paints, finishes, and colors.

Stabilizing deteriorated or damaged wood as a preliminary measure, when necessary, prior to undertaking appropriate preservation work.

Protecting and maintaining wood features by providing proper drainage so that water is not allowed to stand on flat, horizontal surfaces or accumulate in decorative features.

Water damage

Applying chemical preservatives to wood features such as beam ends or outriggers that are exposed to decay hazards and are traditionally unpainted.

Retaining coatings such as paint that help protect the wood from moisture and ultraviolet light. Paint removal should be considered only where there is paint surface deterioration and as part of an overall maintenance program which involves repainting or applying other appropriate protective coatings.

Inspecting painted wood surfaces to determine whether repainting is necessary or if cleaning is all that is required.

Removing damaged or deteriorated paint to the next sound layer using the gentlest method possible (handscraping and handsanding), then repainting.

Not Recommended ✗

Altering wood features which are important in defining the overall historic character of the building so that, as a result, the character is diminished.

Replacing historic wood features instead of repairing or replacing only the deteriorated wood.

Changing the type of paint or finish and its color.

Failing to stabilize deteriorated or damaged wood until additional work is undertaken, thus allowing further damage to occur to the historic building.

Failing to identify, evaluate, and treat the causes of wood deterioration, including faulty flashing, leaking gutters, cracks and holes in siding, deteriorated caulking in joints and seams, plant material growing too close to wood surfaces, or insect or fungus infestation.

Using chemical preservatives such as creosote which, unless they were used historically, can change the appearance of wood features.

Stripping paint or other coatings to reveal bare wood, thus exposing historically coated surfaces to the effects of accelerated weathering.

Removing paint that is firmly adhering to, and thus, protecting wood surfaces.

Using destructive paint removal methods such as propane or butane torches, sandblasting or waterblasting. These methods can irreversibly damage historic woodwork.

Preservation

Recommended

Using with care electric hot-air guns on decorative wood features and electric heat plates on flat wood surfaces when paint is so deteriorated that total removal is necessary prior to repainting.

Using chemical strippers primarily to supplement other methods such as handscraping, handsanding and the above-recommended thermal devices. Detachable wooden elements such as shutters, doors, and columns may—with the proper safeguards—be chemically dip-stripped.

Applying compatible paint coating systems following proper surface preparation.

Not Recommended

Using thermal devices improperly so that the historic woodwork is scorched.

Failing to neutralize the wood thoroughly after using chemicals so that new paint does not adhere.

Allowing detachable wood features to soak too long in a caustic solution so that the wood grain is raised and the surface roughened.

Failing to follow manufacturers' product and application instructions when repainting exterior woodwork.

Maximizing retention of historic materials and features is the primary goal of Preservation as demonstrated here in these "before" and "after" photographs. Aside from some minor repairs and limited replacement of deteriorated material, work on this house consisted primarily of repainting the wood exterior. Photos: Historic Charleston Foundation.

Preservation

Recommended

Repainting with colors that are appropriate to the historic building and district.

Evaluating the existing condition of the wood to determine whether more than protection and maintenance are required, that is, if repairs to wood features will be necessary.

Repairing, stabilizing, and conserving fragile wood using well-tested consolidants, when appropriate. Repairs should be physically and visually compatible and identifiable upon close inspection for future research.

Repairing wood features by patching, piecing-in, or otherwise reinforcing the wood using recognized preservation methods. The new work should be unobtrusively dated to guide future research and treatment.

Not Recommended

Using new colors that are inappropriate to the historic building or district.

Failing to undertake adequate measures to assure the protection of wood features.

Removing wood that could be stabilized and conserved; or using untested consolidants and untrained personnel, thus causing further damage to fragile historic materials.

Removing wood that could be repaired, using improper repair techniques, or failing to document the new work.

The following work is highlighted to indicate that it represents the greatest degree of intervention that is generally recommended within the treatment **Preservation***, and should only be considered after protection, stabilization, and repair concerns have been addressed.*

Recommended

Limited Replacement in Kind

Replacing in kind extensively deteriorated or missing parts of wood features when there are surviving prototypes such as brackets, molding, or sections of siding. New work should match the old in material, design, color, and texture; and be unobtrusively dated to guide future research and treatment.

Not Recommended

Replacing an entire wood feature such as a column or stairway when limited replacement of deteriorated and missing parts is appropriate.

Using replacement material that does not match the historic wood feature; or failing to properly document the new work.

Preservation

Building Exterior

Architectural Metals: Cast iron, steel, pressed tin, copper, aluminum, and zinc

Recommended

Identifying, retaining, and preserving architectural metal features such as columns, capitals, window hoods, or stairways that are important in defining the overall historic character of the building; and their finishes and colors. Identification is also critical to differentiate between metals prior to work. Each metal has unique properties and thus requires different treatments.

Stabilizing deteriorated or damaged architectural metals as a preliminary measure, when necessary, prior to undertaking appropriate preservation work.

Protecting and maintaining architectural metals from corrosion by providing proper drainage so that water does not stand on flat, horizontal surfaces or accumulate in curved, decorative features.

Cleaning architectural metals, when appropriate, to remove corrosion prior to repainting or applying other appropriate protective coatings.

Identifying the particular type of metal prior to any cleaning procedure and then testing to assure that the gentlest cleaning method possible is selected or determining that cleaning is inappropriate for the particular metal.

Not Recommended

Altering architectural metal features which are important in defining the overall historic character of the building so that, as a result, the character is diminished.

Replacing historic metal features instead of repairing or replacing only the deteriorated metal.

Changing the type of finish or its historic color or accent scheme.

Failing to stabilize deteriorated or damaged architectural metals until additional work is undertaken, thus allowing further damage to occur to the historic building.

Failing to identify, evaluate, and treat the causes of corrosion, such as moisture from leaking roofs or gutters.

Placing incompatible metals together without providing a reliable separation material. Such incompatibility can result in galvanic corrosion of the less noble metal, e.g., copper will corrode cast iron, steel, tin, and aluminum.

Exposing metals which were intended to be protected from the environment.

Applying paint or other coatings to metals such as copper, bronze, or stainless steel that were meant to be exposed.

Using cleaning methods which alter or damage the historic color, texture, and finish of the metal; or cleaning when it is inappropriate for the metal.

Removing the patina of historic metal. The patina may be a protective coating on some metals, such as bronze or copper, as well as a significant historic finish.

Preservation

Recommended

Cleaning soft metals such as lead, tin, copper, terneplate, and zinc with appropriate chemical methods because their finishes can be easily abraded by blasting methods.

Using the gentlest cleaning methods for cast iron, wrought iron, and steel—hard metals—in order to remove paint buildup and corrosion. If handscraping and wire brushing have proven ineffective, low pressure grit blasting may be used as long as it does not abrade or damage the surface.

Applying appropriate paint or other coating systems after cleaning in order to decrease the corrosion rate of metals or alloys.

Repainting with colors that are appropriate to the historic building or district.

Applying an appropriate protective coating such as lacquer to an architectural metal feature such as a bronze door which is subject to heavy pedestrian use.

Evaluating the existing condition of the architectural metals to determine whether more than protection and maintenance are required, that is, if repairs to features will be necessary.

Not Recommended

Cleaning soft metals such as lead, tin, copper, terneplate, and zinc with grit blasting which will abrade the surface of the metal.

Failing to employ gentler methods prior to abrasively cleaning cast iron, wrought iron or steel; or using high pressure grit blasting.

Failing to re-apply protective coating systems to metals or alloys that require them after cleaning so that accelerated corrosion occurs.

Using new colors that are inappropriate to the historic building or district.

Failing to assess pedestrian use or new access patterns so that architectural metal features are subject to damage by use or inappropriate maintenance such as salting adjacent sidewalks.

Failing to undertake adequate measures to assure the protection of architectural metal features.

Preservation

Recommended

Repairing, stabilizing, and conserving fragile architectural metals using well-tested consolidants, when appropriate. Repairs should be physically and visually compatible and identifiable upon close inspection for future research.

Repairing architectural metal features by patching, piecing-in, or otherwise reinforcing the metal using recognized preservation methods. The new work should be unobtrusively dated to guide future research and treatment.

Not Recommended

Removing architectural metals that could be stabilized and conserved; or using untested consolidants and untrained personnel, thus causing further damage to fragile historic materials.

Removing architectural metals that could be repaired, using improper repair techniques, or failing to document the new work.

Two examples of "limited replacement in kind" point out an appropriate scope of work within the treatment, Preservation. (a) One metal modillion that has sustained damage from a faulty gutter will need to be replaced; and (b) targeted repairs to deteriorated wood cornice elements (fascia board and modillions) meant that most of the historic materials were retained in the work.

Preservation

The following work is highlighted to indicate that it represents the greatest degree of intervention generally recommended within the treatment **Preservation**, *and should only be considered after protection, stabilization, and repair concerns have been addressed.*

Recommended	*Not Recommended*
Limited Replacement in Kind	
Replacing in kind extensively deteriorated or missing parts of architectural metal features when there are surviving prototypes such as porch balusters, column capitals or bases, or porch cresting. The new work should match the old in material, design, and texture; and be unobtrusively dated to guide future research and treatment.	Replacing an entire architectural metal feature such as a column or balustrade when limited replacement of deteriorated and missing parts is appropriate. Using replacement material that does not match the historic metal feature; or failing to properly document the new work.

Preservation

Building Exterior

Roofs

Recommended

Identifying, retaining, and preserving roofs—and their functional and decorative features—that are important in defining the overall historic character of the building. This includes the roof's shape, such as hipped, gambrel, and mansard; decorative features such as cupolas, cresting, chimneys, and weathervanes; and roofing material such as slate, wood, clay tile, and metal, as well as its size, color, and patterning.

Stabilizing deteriorated or damaged roofs as a preliminary measure, when necessary, prior to undertaking appropriate preservation work.

Not Recommended

Altering the roof and roofing materials which are important in defining the overall historic character of the building so that, as a result, the character is diminished.

Replacing historic roofing material instead of repairing or replacing only the deteriorated material.

Changing the type or color of roofing materials.

Failing to stabilize a deteriorated or damaged roof until additional work is undertaken, thus allowing further damage to occur to the historic building.

It is particularly important to preserve materials that contribute to a building's historic character, such as this highly visible slate roof. In the event that repair and limited replacement are necessary, all new slate would need to match the old exactly. Photo: Jeffrey S. Levine.

Preservation

Recommended

Protecting and maintaining a roof by cleaning the gutters and downspouts and replacing deteriorated flashing. Roof sheathing should also be checked for proper venting to prevent moisture condensation and water penetration; and to insure that materials are free from insect infestation.

Providing adequate anchorage for roofing material to guard against wind damage and moisture penetration.

Protecting a leaking roof with plywood and building paper until it can be properly repaired.

Repairing a roof by reinforcing the historic materials which comprise roof features using recognized preservation methods. The new work should be unobtrusively dated to guide future research and treatment.

Not Recommended

Failing to clean and maintain gutters and downspouts properly so that water and debris collect and cause damage to roof fasteners, sheathing, and the underlying structure.

Allowing roof fasteners, such as nails and clips to corrode so that roofing material is subject to accelerated deterioration.

Permitting a leaking roof to remain unprotected so that accelerated deterioration of historic building materials—masonry, wood, plaster, paint and structural members—occurs.

Removing materials that could be repaired, using improper repair techniques, or failing to document the new work.

Failing to reuse intact slate or tile when only the roofing substrate needs replacement.

The following work is highlighted to indicate that it represents the greatest degree of intervention generally recommended within the treatment **Preservation***, and should only be considered after protection, stabilization, and repair concerns have been addressed.*

Recommended

Limited Replacement in Kind

Replacing in kind extensively deteriorated or missing parts of roof features or roof coverings when there are surviving prototypes such as cupola louvers, dentils, dormer roofing; or slates, tiles, or wood shingles on a main roof. the new work should match the old in material, design, color, and texture; and be unobtrusively dated to guide future research and treatment.

Not Recommended

Replacing an entire roof feature such as a cupola or dormer when limited replacement of deteriorated and missing parts is appropriate.

Using replacement material that does not match the historic roof feature; or failing to properly document the new work.

Preservation

Building Exterior

Windows

Recommended

Identifying, retaining, and preserving windows—and their functional and decorative features—that are important in defining the overall historic character of the building. Such features can include frames, sash, muntins, glazing, sills, heads, hoodmolds, panelled or decorated jambs and moldings, and interior and exterior shutters and blinds.

Not Recommended

Altering windows or window features which are important in defining the historic character of the building so that, as a result, the character is diminished.

Changing the historic appearance of windows by replacing materials, finishes, or colors which noticeably change the sash, depth of reveal, and muntin configuration; the reflectivity and color of the glazing; or the appearance of the frame.

Obscuring historic window trim with metal or other material.

Preserving a building's historic windows generally involves scraping, sanding, and re-painting. While some repair work will most likely be undertaken within the scope of work on this institutional building, replacement of the window units is usually not an appropriate Preservation treatment. Photo: Chuck Fisher.

Preservation

Recommended

Conducting an indepth survey of the condition of existing windows early in preservation planning so that repair and upgrading methods and possible replacement options can be fully explored.

Stabilizing deteriorated or damaged windows as a preliminary measure, when necessary, prior to undertaking appropriate preservation work.

Protecting and maintaining the wood and architectural metals which comprise the window frame, sash, muntins, and surrounds through appropriate surface treatments such as cleaning, rust removal, limited paint removal, and re-application of protective coating systems.

Making windows weathertight by re-caulking and replacing or installing weatherstripping. These actions also improve thermal efficiency.

Evaluating the existing condition of materials to determine whether more than protection and maintenance are required, i.e. if repairs to windows and window features will be required.

Repairing window frames and sash by patching, piecing-in, consolidating or otherwise reinforcing them using recognized preservation methods. The new work should be unobtrusively dated to guide future research and treatment.

Not Recommended

Replacing windows solely because of peeling paint, broken glass, stuck sash, and high air infiltration. These conditions, in themselves, are no indication that windows are beyond repair.

Failing to stabilize a deteriorated or damaged window until additional work is undertaken, thus allowing further damage to occur to the historic building.

Failing to provide adequate protection of materials on a cyclical basis so that deterioration of the window results.

Retrofitting or replacing windows rather than maintaining the sash, frame, and glazing.

Failing to undertake adequate measures to assure the protection of historic windows.

Failing to protect the historic glazing when repairing windows.

Removing material that could be repaired, using improper repair techniques, or failing to document the new work.

Failing to reuse serviceable window hardware such as brass sash lifts and sash locks.

Preservation

The following work is highlighted to indicate that it represents the greatest degree of intervention generally recommended within the treatment **Preservation**, *and should only be considered after protection, stabilization, and repair concerns have been addressed.*

Recommended	*Not Recommended*
Limited Replacement in Kind	
Replacing in kind extensively deteriorated or missing parts of windows when there are surviving prototypes such as frames, sash, sills, glazing, and hoodmolds. The new work should match the old in material, design, color, and texture; and be unobtrusively dated to guide future research and treatment.	Replacing an entire window when limited replacement of deteriorated and missing parts is appropriate. Using replacement material that does not match the historic window; or failing to properly document the new work.

Preservation

Building Exterior

Entrances and Porches

Recommended

Identifying, retaining, and preserving entrances and porches—and their functional and decorative features—that are important in defining the overall historic character of the building such as doors, fanlights, sidelights, pilasters, entablatures, columns, balustrades, and stairs.

Stabilizing deteriorated or damaged entrances and porches as a preliminary measure, when necessary, prior to undertaking appropriate preservation work.

Protecting and maintaining the masonry, wood, and architectural metals that comprise entrances and porches through appropriate surface treatments such as cleaning, rust removal, limited paint removal, and re-application of protective coating systems.

Evaluating the existing condition of materials to determine whether more than protection and maintenance are required, that is, repairs to entrance and porch features will be necessary.

Repairing entrances and porches by reinforcing the historic materials using recognized preservation methods. The new work should be unobtrusively dated to guide future research and treatment.

Not Recommended

Altering entrances and porches which are important in defining the overall historic character of the building so that, as a result, the character is diminished.

Replacing historic entrance and porch features instead of repairing or replacing only the deteriorated material.

Failing to stabilize a deteriorated or damaged entrance or porch until additional work is undertaken, thus allowing further damage to occur to the historic building.

Failing to provide adequate protection to materials on a cyclical basis so that deterioration of entrances and porches results.

Failing to undertake adequate measures to assure the protection of historic entrances and porches.

Removing material that could be repaired, using improper repair techniques, or failing to document the new work.

Preservation

The following work is highlighted to indicate that it represents the greatest degree of intervention generally recommended within the treatment **Preservation**, *and should only be considered after protection, stabilization, and repair concerns have been addressed.*

Limited Replacement in Kind

Recommended

Replacing in kind extensively deteriorated or missing parts of repeated entrance and porch features when there are surviving prototypes such as balustrades, cornices, entablatures, columns, sidelights, and stairs. The new work should match the old in material, design, color, and texture; and be unobtrusively dated to guide future research and treatment.

Not Recommended

Replacing an entire entrance or porch feature when limited replacement of deteriorated and missing parts is appropriate.

Using replacement material that does not match the historic entrance or porch feature; or failing to properly document the new work.

Preservation

Building Exterior

Storefronts

Recommended

Identifying, retaining, and preserving storefronts—and their functional and decorative features—that are important in defining the overall historic character of the building such as display windows, signs, doors, transoms, kick plates, corner posts, and entablatures.

Stabilizing deteriorated or damaged storefronts as a preliminary measure, when necessary, prior to undertaking appropriate preservation work.

Not Recommended

Altering storefronts—and their features—which are important in defining the overall historic character of the building so that, as a result, the character is diminished.

Replacing historic storefront features instead of repairing or replacing only the deteriorated material.

Failing to stabilize a deteriorated or damaged storefront until additional work is undertaken, thus allowing further damage to occur to the historic building.

The original form and features of this 1920s storefront have been retained through Preservation. Photo: David W. Look, AIA.

Preservation

Recommended	*Not Recommended*
Protecting and maintaining masonry, wood, and architectural metals which comprise storefronts through appropriate treatments such as cleaning, rust removal, limited paint removal, and reapplication of protective coating systems.	Failing to provide adequate protection of materials on a cyclical basis so that deterioration of storefront features results.
Protecting storefronts against arson and vandalism before work begins by boarding up windows and doors and installing alarm systems that are keyed into local protection agencies.	Permitting entry into the building through unsecured or broken windows and doors so that interior features and finishes are damaged by exposure to weather or vandalism.
	Stripping storefronts of historic material such as wood, cast iron, terra cotta, carrara glass, and brick.
Evaluating the existing condition of storefront materials to determine whether more than protection and maintenance are required, that is, if repairs to features will be necessary.	Failing to undertake adequate measures to assure the preservation of the historic storefront.
Repairing storefronts by reinforcing the historic materials using recognized preservation methods. The new work should be unobtrusively dated to guide future research and treatment.	Removing material that could be repaired, using improper repair techniques, or failing to document the new work.

The following work is highlighted to indicate that it represents the greatest degree of intervention generally recommended within the treatment **Preservation***, and should only be considered after protection, stabilization, and repair concerns have been addressed.*

Recommended	*Not Recommended*
Limited Replacement in Kind	
Replacing in kind extensively deteriorated or missing parts of storefronts where there are surviving prototypes such as transoms, kick plates, pilasters, or signs. The new work should match the old in materials, design, color, and texture; and be unobtrusively dated to guide future research and treatment.	Replacing an entire storefront when limited replacement of deteriorated and missing parts is appropriate.

Using replacement material that does not match the historic storefront feature; or failing to properly document the new work. |

Preservation

Building Interior

Structural Systems

Recommended

Identifying, retaining, and preserving structural systems—and individual features of systems—that are important in defining the overall historic character of the building, such as post and beam systems, trusses, summer beams, vigas, cast iron columns, above-grade stone foundation walls, or load-bearing brick or stone walls.

Stabilizing deteriorated or damaged structural systems as a preliminary measure, when necessary, prior to undertaking appropriate preservation work.

Protecting and maintaining the structural system by cleaning the roof gutters and downspouts; replacing roof flashing; keeping masonry, wood, and architectural metals in a sound condition; and ensuring that structural members are free from insect infestation.

Examining and evaluating the existing condition of the structural system and its individual features using non-destructive techniques such as X-ray photography.

Not Recommended

Altering visible features of historic structural systems which are important in defining the overall historic character of the building so that, as a result, the character is diminished.

Overloading the existing structural system; or installing equipment or mechanical systems which could damage the structure.

Replacing a loadbearing masonry wall that could be augmented and retained.

Leaving known structural problems untreated such as deflection of beams, cracking and bowing of walls, or racking of structural members.

Utilizing treatments or products that accelerate the deterioration of structural material such as introducing urea-formaldehyde foam insulation into frame walls.

Failing to stabilize a deteriorated or damaged structural system until additional work is undertaken, thus allowing further damage to occur to the historic building.

Failing to provide proper building maintenance so that deterioration of the structural system results. Causes of deterioration include subsurface ground movement, vegetation growing too close to foundation walls, improper grading, fungal rot, and poor interior ventilation that results in condensation.

Utilizing destructive probing techniques that will damage or destroy structural material.

Preservation

Recommended

Repairing the structural system by augmenting or upgrading individual parts or features using recognized preservation methods. For example, weakened structural members such as floor framing can be paired with a new member, braced, or otherwise supplemented and reinforced.

Not Recommended

Upgrading the building structurally in a manner that diminishes the historic character of the exterior, such as installing strapping channels or removing a decorative cornice; or damages interior features or spaces.

Replacing a structural member or other feature of the structural system when it could be augmented and retained.

The following work is highlighted to indicate that it represents the greatest degree of intervention generally recommended within the treatment **Preservation** *and should be only be considered after protection, stabilization, and repair concerns have been addressed.*

Recommended

Limited Replacement in Kind

Replacing in kind those visible portions or features of the structural system that are either extensively deteriorated or missing when there are surviving prototypes such as cast iron columns and sections of loadbearing walls. The new work should match the old in materials, design, color, and texture; and be unobtrusively dated to guide future research and treatment.

Considering the use of substitute material for unexposed structural replacements, such as roof rafters or trusses. Substitute material should, at a minimum, have equal loadbearing capabilities, and be unobtrusively dated to guide future research and treatment.

Not Recommended

Replacing an entire visible feature of the structural system when limited replacement of deteriorated and missing portions is appropriate.

Using material for a portion of an exposed structural feature that does not match the historic feature; or failing to properly document the new work.

Using substitute material that does not equal the loadbearing capabilities of the historic material or design or is otherwise physically or chemically incompatible.

Building Interior *Structural Systems*

Preservation

Building Interior

Spaces, Features, and Finishes

Recommended

Interior Spaces

Identifying, retaining, and preserving a floor plan or interior spaces that are important in defining the overall historic character of the building. This includes the size, configuration, proportion, and relationship of rooms and corridors; the relationship of features to spaces; and the spaces themselves such as lobbies, reception halls, entrance halls, double parlors, theaters, auditoriums, and important industrial or commercial spaces.

Not Recommended

Altering a floor plan or interior spaces—including individual rooms—which are important in defining the overall historic character of the building so that, as a result, the character is diminished.

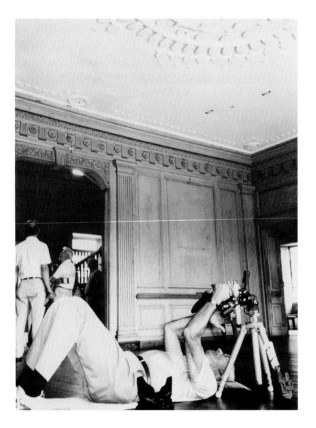

Careful documentation of a building's physical condition is the critical first step in determining an appropriate level of intervention. (a) This may include relating the historical research to existing materials and features; or (b) documenting a particular problem such as this cracked ceiling. Photo (a): Jean E. Travers; Photo (b): Lee H. Nelson, FAIA.

Preservation

Recommended

Interior Features and Finishes

Identifying, retaining, and preserving interior features and finishes that are important in defining the overall historic character of the building, including columns, cornices, baseboards, fireplaces and mantels, panelling, light fixtures, hardware, and flooring; and wallpaper, plaster, paint, and finishes such as stencilling, marbling, and graining; and other decorative materials that accent interior features and provide color, texture, and patterning to walls, floors, and ceilings.

Stabilizing deteriorated or damaged interior features and finishes as a preliminary measure, when necessary, prior to undertaking appropriate preservation work.

Protecting and maintaining masonry, wood, and architectural metals that comprise interior features through appropriate surface treatments such as cleaning, rust removal, limited paint removal, and reapplication of protective coating systems.

Not Recommended

Altering features and finishes which are important in defining the overall historic character of the building so that, as a result, the character is diminished.

Replacing historic interior features and finishes instead of repairing or replacing only the deteriorated masonry.

Installing new decorative material that obscures or damages character-defining interior features or finishes.

Removing historic finishes, such as paint and plaster, or historic wall coverings, such as wallpaper.

Applying paint, plaster, or other finishes to surfaces that have been historically unfinished.

Stripping paint to bare wood rather than repairing or reapplying grained or marbled finishes to features such as doors and paneling.

Changing the type of finish or its color, such as painting a previously varnished wood feature.

Failing to stabilize a deteriorated or damaged interior feature or finish until additional work is undertaken, thus allowing further damage to occur to the historic building.

Failing to provide adequate protection to materials on a cyclical basis so that deterioration of interior features results.

Preservation

Recommended

Protecting interior features and finishes against arson and vandalism before project work begins, boarding-up windows, and installing fire alarm systems that are keyed to local protection agencies.

Protecting interior features such as a staircase, mantel, or decorative finishes and wall coverings against damage during project work by covering them with heavy canvas or plastic sheets.

Installing protective coverings in areas of heavy pedestrian traffic to protect historic features such as wall coverings, parquet flooring and panelling.

Removing damaged or deteriorated paints and finishes to the next sound layer using the gentlest method possible, then repainting or refinishing using compatible paint or other coating systems.

Repainting with colors that are appropriate to the historic building.

Limiting abrasive cleaning methods to certain industrial warehouse buildings where the interior masonry or plaster features do not have distinguishing design, detailing, tooling, or finishes; and where wood features are not finished, molded, beaded, or worked by hand. Abrasive cleaning should only be considered after other, gentler methods have been proven ineffective.

Evaluating the existing condition of materials to determine whether more than protection and maintenance are required, that is, if repairs to interior features and finishes will be necessary.

Not Recommended

Permitting entry into historic buildings through unsecured or broken windows and doors so that the interior features and finishes are damaged by exposure to weather or vandalism.

Stripping interiors of features such as woodwork, doors, windows, light fixtures, copper piping, radiators; or of decorative materials.

Failing to provide proper protection of interior features and finishes during work so that they are gouged, scratched, dented, or otherwise damaged.

Failing to take new use patterns into consideration so that interior features and finishes are damaged.

Using destructive methods such as propane or butane torches or sandblasting to remove paint or other coatings. These methods can irreversibly damage the historic materials that comprise interior features.

Using new paint colors that are inappropriate to the historic building.

Changing the texture and patina of character-defining features through sandblasting or use of abrasive methods to remove paint, discoloration or plaster. This includes both exposed wood (including structural members) and masonry.

Failing to undertake adequate measures to assure the protection of interior features and finishes.

Preservation

Recommended

Repairing historic interior features and finishes by reinforcing the materials using recognized preservation methods. The new work should match the old in material, design, color, and texture; and be unobtrusively dated to guide future research and treatment.

Not Recommended

Removing materials that could be repaired, using improper techniques, or failing to document the new work.

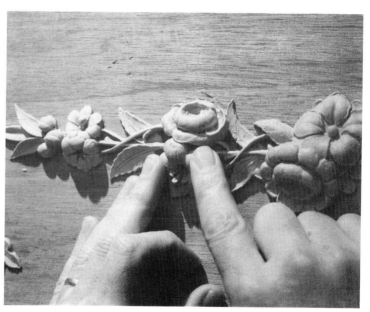

In Preservation, an appropriate level of intervention is established prior to work in order to maximize retention of historic materials.
(a) A conservator is applying adhesive to 19th century composition ornament that has delaminated from its wood substrate.
(b) The compo fragment is carefully held in place until the quick-setting adhesive takes hold. Photos: Jonathan Thornton.

Preservation

The following work is highlighted to indicate that it represents the greatest degree of intervention generally recommended within the treatment **Preservation***, and should only be considered after protection, stabilization, and repair concerns have been addressed.*

Recommended

Limited Replacement in Kind

Replacing in kind extensively deteriorated or missing parts of repeated interior features when there are surviving prototypes such as stairs, balustrades, wood panelling, columns; or decorative wall coverings or ornamental tin or plaster ceilings. New work should match the old in material, design, color, and texture; and be unobtrusively dated to guide future research and treatment.

Not Recommended

Replacing an entire interior feature when limited replacement of deteriorated and missing parts is appropriate.

Using replacement material that does not match the interior feature; or failing to properly document the new work.

Building Interior

Mechanical Systems: Heating, Air Conditioning, Electrical, and Plumbing

Recommended

Identifying, retaining, and preserving visible features of early mechanical systems that are important in defining the overall historic character of the building, such as radiators, vents, fans, grilles, plumbing fixtures, switchplates, and lights.

Stabilizing deteriorated or damaged mechanical systems as a preliminary measure, when necessary, prior to undertaking appropriate preservation work.

Protecting and maintaining mechanical, plumbing, and electrical systems and their features through cyclical cleaning and other appropriate measures.

Preventing accelerated deterioration of mechanical systems by providing adequate ventilation of attics, crawlspaces, and cellars so that moisture problems are avoided.

Improving the energy efficiency of existing mechanical systems to help reduce the need for elaborate new equipment.

Repairing mechanical systems by augmenting or upgrading system parts, such as installing new pipes and ducts; rewiring; or adding new compressors or boilers.

Replacing in kind those visible features of mechanical systems that are either extensively deteriorated or are prototypes such as ceiling fans, switchplates, radiators, grilles, or plumbing fixtures.

Not Recommended

Removing or altering visible features of mechanical systems that are important in defining the overall historic character of the building so that, as a result, the character is diminished.

Failing to stabilize a deteriorated or damaged mechanical system until additional work is undertaken, thus allowing further damage to occur to the historic building.

Failing to provide adequate protection of materials on a cyclical basis so that deterioration of mechanical systems and their visible features results.

Enclosing mechanical systems in areas that are not adequately ventilated so that deterioration of the systems results.

Installing unnecessary climate control systems which can add excessive moisture to the building. This additional moisture can either condense inside, damaging interior surfaces, or pass through interior walls to the exterior, potentially damaging adjacent materials as it migrates.

Replacing a mechanical system or its functional parts when it could be upgraded and retained.

Installing a visible replacement feature that does not convey the same visual appearance.

Preservation

The following should be considered in a **Preservation** *project when the installation of new mechanical equipment or system is required to make the building functional.*

Recommended	*Not Recommended*
Installing a new mechanical system if required, so that it causes the least alteration possible to the building.	Installing a new mechanical system so that character-defining structural or interior features are radically changed, damaged, or destroyed.
Providing adequate structural support for new mechanical equipment.	Failing to consider the weight and design of new mechanical equipment so that, as a result, historic structural members or finished surfaces are weakened or cracked.
Installing the vertical runs of ducts, pipes, and cables in closets, service rooms, and wall cavities.	Installing vertical runs of ducts, pipes, and cables in places where they will obscure character-defining features.
	Concealing mechanical equipment in walls or ceilings in a manner that requires excessive removal of historic building material.
Installing air conditioning in such a manner that historic features are not damaged or obscured and excessive moisture is not generated that will accelerate deterioration of historic materials.	Cutting through features such as masonry walls in order to install air conditioning units.

Preservation

Building Site

Recommended

Identifying, retaining, and preserving buildings and their features as well as features of the site that are important in defining its overall historic character. Site features may include circulation systems such as walks, paths, roads, or parking; vegetation such as trees, shrubs, fields, or herbaceous plant material; landforms such as terracing, berms or grading; furnishings such as lights, fences, or benches; decorative elements such as sculpture, statuary or monuments; water features including fountains, streams, pools, or lakes; and subsurface archeological features which are important in defining the history of the site.

Retaining the historic relationship between buildings and the landscape.

Stabilizing deteriorated or damaged building and site features as a preliminary measure, when necessary, prior to undertaking appropriate preservation work.

Not Recommended

Altering buildings and their features or site features which are important in defining the overall historic character of the property so that, as a result, the character is diminished.

Removing or relocating buildings or landscape features, thus destroying the historic relationship between buildings and the landscape.

Failing to stabilize a deteriorated or damaged building or site feature until additional work is undertaken, thus allowing further damage to occur to the building site.

Drayton Hall, near Charleston, South Carolina, is an excellent example of an evolved 18th century plantation. Of particular note in this photograph are the landscape features added in the late 19th century—a reflecting pond and rose mound. With an overall Preservation treatment plan, these later features have been retained and protected. If a Restoration treatment had been selected, later features of the landscape as well as changes to the house would have been removed. Photo: Courtesy, National Trust for Historic Preservation.

Building Site 51

Preservation

Recommended	*Not Recommended*

Protecting and maintaining buildings and sites by providing proper drainage to assure that water does not erode foundation walls; drain toward the building; or damage or erode the landscape.

Failing to maintain adequate site drainage so that buildings and site features are damaged or destroyed; or alternatively, changing the site grading so that water no longer drains properly.

Minimizing disturbance of terrain around buildings or elsewhere on the site, thus reducing the possibility of destroying or damaging important landscape features or archeological resources.

Introducing heavy machinery into areas where it may disturb or damage important landscape features or archeological resources.

Surveying and documenting areas where the terrain will be altered to determine the potential impact to important landscape features or archeological resources.

Failing to survey the building site prior to beginning work which results in damage to, or destruction of, important landscape features or archeological resources.

Protecting, e.g., preserving in place, important archeological resources.

Leaving known archeological material unprotected so that it is damaged during preservation work.

Planning and carrying out any necessary investigation using professional archeologists and modern archeological methods when preservation in place is not feasible.

Permitting unqualified personnel to perform data recovery on archeological resources so that improper methodology results in the loss of important archeological material.

Preserving important landscape features, including ongoing maintenance of historic plant material.

Allowing important landscape features to be lost or damaged due to a lack of maintenance.

Protecting building and landscape features against arson and vandalism before preservation work begins, i.e., erecting protective fencing and installing alarm systems that are keyed into local protection agencies.

Permitting the property to remain unprotected so that the building and landscape features or archeological resources are damaged or destroyed.

Removing or destroying features from the buildings or site such as wood siding, iron fencing, masonry balustrades, or plant material.

Providing continued protection of historic building materials and plant features through appropriate cleaning, rust removal, limited paint removal, and re-application of protective coating systems; and pruning and vegetation management.

Failing to provide adequate protection of materials on a cyclical basis so that deterioration of building and site feature results.

Preservation

Recommended

Evaluating the existing condition of materials and features to determine whether more than protection and maintenance are required, that is, if repairs to building and site features will be necessary.

Repairing features of the building and site by reinforcing historic materials using recognized preservation methods. The new work should be unobtrusively dated to guide future research and treatment.

Not Recommended

Failing to undertake adequate measures to assure the protection of building and site features.

Removing materials that could be repaired, using improper repair techniques, or failing to document the new work.

The following work is highlighted to indicate that it represents the greatest degree of intervention generally recommended within the treatment **Preservation***, and should only be considered after protection, stabilization, and repair concerns have been addressed.*

Recommended

Limited Replacement in Kind

Replacing in kind extensively deteriorated or missing parts of the building or site where there are surviving prototypes such as part of a fountain, or portions of a walkway. New work should match the old in materials, design, color, and texture; and be unobtrusively dated to guide future research and treatment.

Not Recommended

Replacing an entire feature of the building or site when limited replacement of deteriorated and missing parts is appropriate.

Using replacement material that does not match the building site feature; or failing to properly document the new work.

Preservation

Setting (District/Neighborhood)

Recommended

Identifying retaining, and preserving building and landscape features which are important in defining the historic character of the setting. Such features can include roads and streets, furnishings such as lights or benches, vegetation, gardens and yards, adjacent open space such as fields, parks, commons or woodlands, and important views or visual relationships.

Retaining the historic relationship between buildings and landscape features of the setting. For example, preserving the relationship between a town common and its adjacent historic houses, municipal buildings, historic roads, and landscape features.

Stabilizing deteriorated or damaged building and landscape features of the setting as a preliminary measure, when necessary, prior to undertaking appropriate preservation work.

Protecting and maintaining historic building materials and plant features through appropriate cleaning, rust removal, limited paint removal, and reapplication of protective coating systems; and pruning and vegetation management.

Protecting building and landscape features against arson and vandalism before preservation work begins by erecting protective fencing and installing alarm systems that are keyed into local preservation agencies.

Evaluating the existing condition of the building and landscape features to determine whether more than protection and maintenance are required, that is, if repairs to features will be necessary.

Not Recommended

Altering those features of the setting which are important in defining the historic character.

Altering the relationship between the buildings and landscape features within the setting by widening existing streets, changing landscape materials, or constructing inappropriately located new streets or parking.

Removing or relocating historic buildings or landscape features, thus destroying their historic relationship within the setting.

Failing to stabilize a deteriorated or damaged building or landscape feature of the setting until additional work is undertaken, thus allowing further damage to the setting to occur.

Failing to provide adequate protection of materials on a cyclical basis which results in the deterioration of building and landscape features.

Permitting the building and setting to remain unprotected so that interior or exterior features are damaged.

Stripping or removing features from buildings or the setting such as wood siding, iron fencing, terra cotta balusters, or plant material.

Failing to undertake adequate measures to assure the protection of building and landscape features.

Preservation

Recommended

Repairing features of the building and landscape using recognized preservation methods. The new work should be unobtrusively dated to guide future research and treatment.

Not Recommended

Removing material that could be repaired, using improper repair techniques, or failing to document the new work.

The following work is highlighted because it represents the greatest degree of intervention generally recommended within the treatment **Preservation***, and should only be considered after protection, stabilization, and repair concerns have been addressed.*

Recommended

Limited Replacement in Kind

Replacing in kind extensively deteriorated or missing parts of building and landscape features where there are surviving prototypes such as porch balustrades or paving materials.

Not Recommended

Replacing an entire feature of the building or landscape when limited replacement of deteriorated and missing parts is appropriate.

Using replacement material that does not match the building or landscape feature; or failing to properly document the new work.

The goal of Preservation is to retain the historic form, materials, and features of the building and its site as they have changed—or evolved—over time. This bank barn was built in the 1820s, then enlarged in 1898 and again in 1914. Today, it continues its role as a working farm structure as a result of sensitive preservation work. This included foundation re-grading; a new gutter system; structural strengthening; and replacement of a severely deteriorated metal roof. Photo: Jack E. Boucher, HABS.

Preservation

Although the work in the following sections is quite often an important aspect of preservation projects, it is usually not part of the overall process of preserving character-defining features (maintenance, repair, and limited replacement); rather, such work is assessed for its potential negative impact on the building's historic character. For this reason, particular care must be taken not to obscure, alter, or damage character-defining features in the process of preservation work.

Energy Efficiency

Recommended	*Not Recommended*
Masonry/Wood/Architectural Metals	
Installing thermal insulation in attics and in unheated cellars and crawlspaces to increase the efficiency of the existing mechanical systems.	Applying thermal insulation with a high moisture content in wall cavities which may damage historic fabric.
Installing insulating material on the inside of masonry walls to increase energy efficiency where there is no character-defining interior molding around the windows or other interior architectural detailing.	Installing wall insulation without considering its effect on interior molding or other architectural detailing.
Windows	
Utilizing the inherent energy conserving features of a building by maintaining windows and louvered blinds in good operable condition for natural ventilation.	Removing historic shading devices rather than keeping them in an operable condition.
Improving thermal efficiency with weatherstripping, storm windows, caulking, interior shades, and if historically appropriate, blinds and awnings.	Replacing historic multi-paned sash with new thermal sash utilizing false muntins.
Installing interior storm windows with air-tight gaskets, ventilating holes, and/or removable clips to insure proper maintenance and to avoid condensation damage to historic windows.	Installing interior storm windows that allow moisture to accumulate and damage the window.
Installing exterior storm windows which do not damage or obscure the windows and frames.	Installing new exterior storm windows which are inappropriate in size or color.
	Replacing windows or transoms with fixed thermal glazing or permitting windows and transoms to remain inoperable rather than utilizing them for their energy conserving potential.

Recommended	*Not Recommended*

Entrances and Porches

Maintaining porches and double vestibule entrances so that they can retain heat or block the sun and provide natural ventilation.	Changing the historic appearance of the building by enclosing porches.

Interior Features

Retaining historic interior shutters and transoms for their inherent energy conserving features.	Removing historic interior features which play an energy conserving role.

Mechanical Systems

Improving energy efficiency of existing mechanical systems by installing insulation in attics and basements.	Replacing existing mechanical systems that could be repaired for continued use.

Building Site

Retaining plant materials, trees, and landscape features which perform passive solar energy functions such as sun shading and wind breaks.	Removing plant materials, trees, and landscape features that perform passive solar energy functions.

Setting (District/Neighborhood)

Maintaining those existing landscape features which moderate the effects of the climate on the setting such as deciduous trees, evergreen wind-blocks, and lakes or ponds.	Stripping the setting of landscape features and landforms so that the effects of wind, rain, and sun result in accelerated deterioration of the historic building.

Preservation

Accessibility Considerations

Recommended

Identifying the historic building's character-defining spaces, features, and finishes so that accessibility code-required work will not result in their damage or loss.

Complying with barrier-free access requirements, in such a manner that character-defining spaces, features, and finishes are preserved.

Working with local disability groups, access specialists, and historic preservation specialists to determine the most appropriate solution to access problems.

Providing barrier-free access that promotes independence for the disabled person to the highest degree practicable, while preserving significant historic features.

Finding solutions to meet accessibility requirements that minimize the impact on the historic building and its site, such as compatible ramps, paths, and lifts.

Not Recommended

Undertaking code-required alterations before identifying those spaces, features, or finishes which are character-defining and must therefore be preserved.

Altering, damaging, or destroying character-defining features in attempting to comply with accessibility requirements.

Making changes to buildings without first seeking expert advice from access specialists and historic preservationists to determine solutions.

Making access modifications that do not provide a reasonable balance between independent, safe access and preservation of historic features.

Making modifications for accessibility without considering the impact on the historic building and its site.

Preservation

Health and Safety Considerations

Recommended

Identifying the historic building's character-defining spaces, features, and finishes so that code-required work will not result in their damage or loss.

Complying with health and safety codes, including seismic code requirements, in such a manner that character-defining spaces, features, and finishes are preserved.

Removing toxic building materials only after thorough testing has been conducted and only after less invasive abatement methods have been shown to be inadequate.

Providing workers with appropriate personal protective equipment for hazards found in the worksite.

Working with local code officials to investigate systems, methods, or devices of equivalent or superior effectiveness and safety to those prescribed by code so that unnecessary alterations can be avoided.

Upgrading historic stairways and elevators to meet health and safety codes in a manner that assures their preservation, i.e., so that they are not damaged or obscured.

Installing sensitively designed fire suppression systems, such as sprinkler systems that result in retention of historic features and finishes.

Applying fire-retardant coatings, such as intumescent paints, which expand during fire to add thermal protection to steel.

Adding a new stairway or elevator to meet health and safety codes in a manner that preserves adjacent character-defining features and spaces.

Not Recommended

Undertaking code-required alterations to a building or site before identifying those spaces, features, or finishes which are character-defining and must therefore be preserved.

Altering, damaging, or destroying character-defining spaces, features, and finishes while making modifications to a building or site to comply with safety codes.

Destroying historic interior features and finishes without careful testing and without considering less invasive abatement methods.

Removing unhealthful building materials without regard to personal and environmental safety.

Making changes to historic buildings without first exploring equivalent health and safety systems, methods, or devices that may be less damaging to historic spaces, features, and finishes.

Damaging or obscuring historic stairways and elevators or altering adjacent spaces in the process of doing work to meet code requirements.

Covering character-defining wood features with fire-resistant sheathing which results in altering their visual appearance.

Using fire-retardant coatings if they damage or obscure character-defining features.

Radically changing, damaging, or destroying character-defining spaces, features, or finishes when adding a new code-required stairway or elevator.

Standards for Rehabilitation & Guidelines for Rehabilitating Historic Buildings

Rehabilitation *is defined as the act or process of making possible a compatible use for a property through repair, alterations, and additions while preserving those portions or features which convey its historical, cultural, or architectural values.*

Standards for Rehabilitation

1. A property will be used as it was historically or be given a new use that requires minimal change to its distinctive materials, features, spaces, and spatial relationships.
2. The historic character of a property will be retained and preserved. The removal of distinctive materials or alteration of features, spaces, and spatial relationships that characterize a property will be avoided.

3. Each property will be recognized as a physical record of its time, place, and use. Changes that create a false sense of historical development, such as adding conjectural features or elements from other historic properties, will not be undertaken.

4. Changes to a property that have acquired historic significance in their own right will be retained and preserved.

5. Distinctive materials, features, finishes, and construction techniques or examples of craftsmanship that characterize a property will be preserved.

6. Deteriorated historic features will be repaired rather than replaced. Where the severity of deterioration requires replacement of a distinctive feature, the new feature will match the old in design, color, texture, and, where possible, materials. Replacement of missing features will be substantiated by documentary and physical evidence.

7. Chemical or physical treatments, if appropriate, will be undertaken using the gentlest means possible. Treatments that cause damage to historic materials will not be used.

8. Archeological resources will be protected and preserved in place. If such resources must be disturbed, mitigation measures will be undertaken.

9. New additions, exterior alterations, or related new construction will not destroy historic materials, features, and spatial relationships that characterize the property. The new work shall be differentiated from the old and will be compatible with the historic materials, features, size, scale and proportion, and massing to protect the integrity of the property and its environment.

10. New additions and adjacent or related new construction will be undertaken in a such a manner that, if removed in the future, the essential form and integrity of the historic property and its environment would be unimpaired.

Guidelines for Rehabilitating Historic Buildings

Introduction

In **Rehabilitation**, historic building materials and character-defining features are protected and maintained as they are in the treatment Preservation; however, an assumption is made prior to work that existing historic fabric has become damaged or deteriorated over time and, as a result, more repair and replacement will be required. Thus, latitude is given in the **Standards for Rehabilitation and Guidelines for Rehabilitation** to replace extensively deteriorated, damaged, or missing features using either traditional or substitute materials. Of the four treatments, only Rehabilitation includes an opportunity to make possible an efficient contemporary use through alterations and additions.

Identify, Retain, and Preserve Historic Materials and Features

Like Preservation, guidance for the treatment **Rehabilitation** begins with recommendations to identify the form and detailing of those architectural materials and features that are important in defining the building's historic character and which must be retained in order to preserve that character. Therefore, guidance on *identifying, retaining, and preserving* character-defining features is always given first. The character of a historic building may be defined by the form and detailing of exterior materials, such as masonry, wood, and metal; exterior features, such as roofs, porches, and windows; interior materials, such as plaster and paint; and interior features, such as moldings and stairways, room configuration and spatial relationships, as well as structural and mechanical systems.

Protect and Maintain Historic Materials and Features

After identifying those materials and features that are important and must be retained in the process of **Rehabilitation** work, then *protecting and maintaining* them are addressed. Protection generally involves the least degree of intervention and is preparatory to other work. For example, protection includes the maintenance of historic material through treatments such as rust removal, caulking, limited paint removal, and re-application of protective coatings; the cyclical cleaning of roof gutter systems; or installation of fencing, alarm systems and other temporary protective measures. Although a historic building will usually require more extensive work, an overall evaluation of its physical condition should always begin at this level.

Repair Historic Materials and Features

Next, when the physical condition of character-defining materials and features warrants additional work *repairing* is recommended. **Rehabilitation** guidance for the repair of historic materials such as masonry, wood, and architectural metals again begins with the least degree of intervention possible such as patching, piecing-in, splicing, consolidating, or otherwise reinforcing or upgrading them according to recognized preservation methods. Repairing also includes the limited replacement in kind—or with compatible substitute material—of extensively deteriorated or missing parts of features when there are

---

Note: The Guidelines for Rehabilitating Historic Buildings in this chapter have already appeared in *The Secretary of the Interior's Standards for Rehabilitation & Illustrated Guidelines for Rehabilitating Historic Buildings*, published in 1992.

Originally built as single-family, semi-detached duplexes, these houses were rehabilitated for a new use as rental apartments. While some alteration to non-significant interior features and spaces was necessary in each one, the exteriors were essentially preserved. Photos: Mistick, Inc.

surviving prototypes (for example, brackets, dentils, steps, plaster, or portions of slate or tile roofing). Although using the same kind of material is always the preferred option, substitute material is acceptable if the form and design as well as the substitute material itself convey the visual appearance of the remaining parts of the feature and finish.

Replace Deteriorated Historic Materials and Features

Following repair in the hierarchy, **Rehabilitation** guidance is provided for *replacing* an entire character-defining feature with new material because the level of deterioration or damage of materials precludes repair (for example, an exterior cornice; an interior staircase; or a complete porch or storefront). If the essential form and detailing are still evident so that the physical evidence can be used to re-establish the feature as an integral part of the rehabilitation, then its replacement is appropriate. Like the guidance for repair, the preferred option is always replacement of the entire feature in kind, that is, with the same material. Because this approach may not always be technically or economically feasible, provisions are made to consider the use of a compatible substitute material.

It should be noted that, while the National Park Service guidelines recommend the replacement of an entire character-defining feature that is extensively deteriorated, they never recommend removal and replacement with new material of a feature that—although damaged or deteriorated—could reasonably be repaired and thus preserved.

Design for the Replacement of Missing Historic Features

When an entire interior or exterior feature is missing (for example, an entrance, or cast iron facade; or a principal staircase), it no longer plays a role in physically defining the historic character of the building unless it can be accurately recovered in form and detailing through the process of carefully documenting the historical appearance. Although accepting the loss is one possibility, where an important architectural feature is missing, its replacement is always recommended in the **Rehabilitation** guidelines as the *first* or preferred, course of action. Thus, if adequate historical, pictorial, and physical documentation exists so that the feature may be accurately reproduced, and if it is desirable to re-establish the feature as part of the building's historical appearance, then designing and constructing a new feature based on such information is appropriate. However, a *second* acceptable option for the replacement feature is a new design that is compatible with the remaining character-defining features of the historic building. The new design should always take into account the size, scale, and material of the historic building itself and, most importantly, should be clearly differentiated so that a false historical appearance is not created.

Alterations/Additions for the New Use

Some exterior and interior alterations to a historic building are generally needed to assure its continued use, but it is most important that such alterations do not radically change, obscure, or destroy character-defining spaces, materials, features, or finishes. Alterations may include providing additional parking space on an existing historic building site; cutting new entrances or windows on secondary elevations; inserting an additional floor; installing an entirely new mechanical system; or creating an atrium or light well. Alteration may also include the selective removal of buildings or other features of the environment or building site that are intrusive and therefore detract from the overall historic character.

The construction of an exterior addition on a historic building may seem to be essential for the new use, but it is emphasized in the **Rehabilitation** guidelines that such new additions should be avoided, if possible, and considered *only* after it is determined that those needs cannot be met by altering secondary, i.e., non character-defining interior spaces. If, after a thorough evaluation of interior solutions, an exterior addition is still judged to be the only viable alterative, it should be designed and constructed to be clearly differentiated from the historic building and so that the character-defining features are not radically changed, obscured, damaged, or destroyed.

Additions and alterations to historic buildings are referenced within specific sections of the **Rehabilitation** guidelines such as Site, Roofs, Structural Systems, etc., but are addressed in detail in *New Additions to Historic Buildings*, found at the end of this chapter.

Energy Efficiency/Accessibility Considerations/Health and Safety Code Considerations

These sections of the guidance address work done to meet accessibility requirements and health and safety code requirements; or retrofitting measures to improve energy efficiency. Although this work is quite often an important aspect of **Rehabilitation** projects, it is usually not a part of the overall process of protecting or repairing character-defining features; rather, such work is assessed for its potential negative impact on the building's historic character. For this reason, particular care must be taken not to radically change, obscure, damage, or destroy character-defining materials or features in the process of meeting code and energy requirements.

***Rehabilitation as a Treatment** When repair and replacement of deteriorated features are necessary; when alterations or additions to the property are planned for a new or continued use; and when its depiction at a particular time is not appropriate, Rehabilitation may be considered as a treatment. Prior to undertaking work, a documentation plan for Rehabilitation should be developed.*

Rehabilitation

Building Exterior

Masonry: Brick, stone, terra cotta, concrete, adobe, stucco and mortar

Recommended

Identifying, retaining, and preserving masonry features that are important in defining the overall historic character of the building such as walls, brackets, railings, cornices, window architraves, door pediments, steps, and columns; and details such as tooling and bonding patterns, coatings, and color.

Protecting and maintaining masonry by providing proper drainage so that water does not stand on flat, horizontal surfaces or accumulate in curved decorative features.

Cleaning masonry only when necessary to halt deterioration or remove heavy soiling.

Carrying out masonry surface cleaning tests after it has been determined that such cleaning is appropriate. Tests should be observed over a sufficient period of time so that both the immediate and the long range effects are known to enable selection of the gentlest method possible.

Not Recommended

Removing or radically changing masonry features which are important in defining the overall historic character of the building so that, as a result, the character is diminished.

Replacing or rebuilding a major portion of exterior masonry walls that could be repaired so that, as a result, the building is no longer historic and is essentially new construction.

Applying paint or other coatings such as stucco to masonry that has been historically unpainted or uncoated to create a new appearance.

Removing paint from historically painted masonry.

Radically changing the type of paint or coating or its color.

Failing to evaluate and treat the various causes of mortar joint deterioration such as leaking roofs or gutters, differential settlement of the building, capillary action, or extreme weather exposure.

Cleaning masonry surfaces when they are not heavily soiled to create a new appearance, thus needlessly introducing chemicals or moisture into historic materials.

Cleaning masonry surfaces without testing or without sufficient time for the testing results to be of value.

Rehabilitation

Recommended	*Not Recommended*
Cleaning masonry surfaces with the gentlest method possible, such as low pressure water and detergents, using natural bristle brushes.	Sandblasting brick or stone surfaces using dry or wet grit or other abrasives. These methods of cleaning permanently erode the surface of the material and accelerate deterioration.
	Using a cleaning method that involves water or liquid chemical solutions when there is any possibility of freezing temperatures.
	Cleaning with chemical products that will damage masonry, such as using acid on limestone or marble, or leaving chemicals on masonry surfaces.
	Applying high pressure water cleaning methods that will damage historic masonry and the mortar joints.
Inspecting painted masonry surfaces to determine whether repainting is necessary.	Removing paint that is firmly adhering to, and thus protecting, masonry surfaces.
Removing damaged or deteriorated paint only to the next sound layer using the gentlest method possible (e.g., hand-scraping) prior to repainting.	Using methods of removing paint which are destructive to masonry, such as sandblasting, application of caustic solutions, or high pressure waterblasting.
Applying compatible paint coating systems following proper surface preparation.	Failing to follow manufacturers' product and application instructions when repainting masonry.
Repainting with colors that are historically appropriate to the building and district.	Using new paint colors that are inappropriate to the historic building and district.
Evaluating the overall condition of the masonry to determine whether more than protection and maintenance are required, that is, if repairs to masonry features will be necessary.	Failing to undertake adequate measures to assure the protection of masonry features.
Repairing masonry walls and other masonry features by repointing the mortar joints where there is evidence of deterioration such as disintegrating mortar, cracks in mortar joints, loose bricks, damp walls, or damaged plasterwork.	Removing nondeteriorated mortar from sound joints, then repointing the entire building to achieve a uniform appearance.
Removing deteriorated mortar by carefully hand-raking the joints to avoid damaging the masonry.	Using electric saws and hammers rather than hand tools to remove deteriorated mortar from joints prior to repointing.

Rehabilitation

Recommended	*Not Recommended*
Duplicating old mortar in strength, composition, color, and texture.	Repointing with mortar of high portland cement content (unless it is the content of the historic mortar). This can often create a bond that is stronger than the historic material and can cause damage as a result of the differing coefficient of expansion and the differing porosity of the material and the mortar.
	Repointing with a synthetic caulking compound.
	Using a "scrub" coating technique to repoint instead of traditional repointing methods.
Duplicating old mortar joints in width and in joint profile.	Changing the width or joint profile when repointing.
Repairing stucco by removing the damaged material and patching with new stucco that duplicates the old in strength, composition, color, and texture.	Removing sound stucco; or repairing with new stucco that is stronger than the historic material or does not convey the same visual appearance.
Using mud plaster as a surface coating over unfired, unstabilized adobe because the mud plaster will bond to the adobe.	Applying cement stucco to unfired, unstabilized adobe. Because the cement stucco will not bond properly, moisture can become entrapped between materials, resulting in accelerated deterioration of the adobe.
Cutting damaged concrete back to remove the source of deterioration (often corrosion on metal reinforcement bars). The new patch must be applied carefully so it will bond satisfactorily with, and match, the historic concrete.	Patching concrete without removing the source of deterioration.
Repairing masonry features by patching, piecing-in, or consolidating the masonry using recognized preservation methods. Repair may also include the limited replacement in kind—or with compatible substitute material—of those extensively deteriorated or missing parts of masonry features when there are surviving prototypes such as terra-cotta brackets or stone balusters.	Replacing an entire masonry feature such as a cornice or balustrade when repair of the masonry and limited replacement of deteriorated or missing parts are appropriate.
	Using a substitute material for the replacement part that does not convey the visual appearance of the surviving parts of the masonry feature or that is physically or chemically incompatible.

Building Exterior *Masonry*

Rehabilitation

Recommended

Applying new or non-historic surface treatments such as water-repellent coatings to masonry only after repointing and only if masonry repairs have failed to arrest water penetration problems.

Replacing in kind an entire masonry feature that is too deteriorated to repair—if the overall form and detailing are still evident—using the physical evidence as a model to reproduce the feature. Examples can include large sections of a wall, a cornice, balustrade, column, or stairway. If using the same kind of material is not technically or economically feasible, then a compatible substitute material may be considered.

Not Recommended

Applying waterproof, water repellent, or non-historic coatings such as stucco to masonry as a substitute for repointing and masonry repairs. Coatings are frequently unnecessary, expensive, and may change the appearance of historic masonry as well as accelerate its deterioration.

Removing a masonry feature that is unrepairable and not replacing it; or replacing it with a new feature that does not convey the same visual appearance.

The following work is highlighted to indicate that it represents the particularly complex technical or design aspects of **Rehabilitation** *projects and should only be considered after the preservation concerns listed above have been addressed.*

Recommended

Design for the Replacement of Missing Historic Features

Designing and installing a new masonry feature such as steps or a door pediment when the historic feature is completely missing. It may be an accurate restoration using historical, pictorial, and physical documentation; or be a new design that is compatible with the size, scale, material, and color of the historic building.

Not Recommended

Creating a false historical appearance because the replaced masonry feature is based on insufficient historical, pictorial, and physical documentation.

Introducing a new masonry feature that is incompatible in size, scale, material and color.

Rehabilitation

Building Exterior
Wood: Clapboard, weatherboard, shingles, and other wooden siding and decorative elements

Recommended

Identifying, retaining, and preserving wood features that are important in defining the overall historic character of the building such as siding, cornices, brackets, window architraves, and doorway pediments; and their paints, finishes, and colors.

Protecting and maintaining wood features by providing proper drainage so that water is not allowed to stand on flat, horizontal surfaces or accumulate in decorative features.

Applying chemical preservatives to wood features such as beam ends or outriggers that are exposed to decay hazards and are traditionally unpainted.

Retaining coatings such as paint that help protect the wood from moisture and ultraviolet light. Paint removal should be considered only where there is paint surface deterioration and as part of an overall maintenance program which involves repainting or applying other appropriate protective coatings.

Not Recommended

Removing or radically changing wood features which are important in defining the overall historic character of the building so that, as a result, the character is diminished.

Removing a major portion of the historic wood from a facade instead of repairing or replacing only the deteriorated wood, then reconstructing the facade with new material in order to achieve a uniform or "improved" appearance.

Radically changing the type of finish or its color or accent scheme so that the historic character of the exterior is diminished.

Stripping historically painted surfaces to bare wood, then applying clear finishes or stains in order to create a "natural look."

Stripping paint or varnish to bare wood rather than repairing or reapplying a special finish, i.e., a grained finish to an exterior wood feature such as a front door.

Failing to identify, evaluate, and treat the causes of wood deterioration, including faulty flashing, leaking gutters, cracks and holes in siding, deteriorated caulking in joints and seams, plant material growing too close to wood surfaces, or insect or fungus infestation.

Using chemical preservatives such as creosote which, unless they were used historically, can change the appearance of wood features.

Stripping paint or other coatings to reveal bare wood, thus exposing historically coated surfaces to the effects of accelerated weathering.

Building Exterior *Wood*

Rehabilitation

Recommended

Inspecting painted wood surfaces to determine whether repainting is necessary or if cleaning is all that is required.

Removing damaged or deteriorated paint to the next sound layer using the gentlest method possible (handscraping and handsanding), then repainting.

Using with care electric hot-air guns on decorative wood features and electric heat plates on flat wood surfaces when paint is so deteriorated that total removal is necessary prior to repainting.

Not Recommended

Removing paint that is firmly adhering to, and thus, protecting wood surfaces.

Using destructive paint removal methods such as propane or butane torches, sandblasting or waterblasting. These methods can irreversibly damage historic woodwork.

Using thermal devices improperly so that the historic woodwork is scorched.

According to the Standards for Rehabilitation, existing historic materials should be protected, maintained and repaired. In an exemplary project, the windows and shutters of this historic residence were carefully preserved.

Building Exterior *Wood*

Rehabilitation

Recommended

Using chemical strippers primarily to supplement other methods such as handscraping, handsanding and the above-recommended thermal devices. Detachable wooden elements such as shutters, doors, and columns may—with the proper safeguards—be chemically dip-stripped.

Applying compatible paint coating systems following proper surface preparation.

Repainting with colors that are appropriate to the historic building and district.

Evaluating the overall condition of the wood to determine whether more than protection and maintenance are required, that is, if repairs to wood features will be necessary.

Repairing wood features by patching, piecing-in, consolidating, or otherwise reinforcing the wood using recognized preservation methods. Repair may also include the limited replacement in kind—or with compatible substitute material—of those extensively deteriorated or missing parts of features where there are surviving prototypes such as brackets, molding, or sections of siding.

Replacing in kind an entire wood feature that is too deteriorated to repair—if the overall form and detailing are still evident—using the physical evidence as a model to reproduce the feature. Examples of wood features include a cornice, entablature or balustrade. If using the same kind of material is not technically or economically feasible, then a compatible substitute material may be considered.

Not Recommended

Failing to neutralize the wood thoroughly after using chemicals so that new paint does not adhere.

Allowing detachable wood features to soak too long in a caustic solution so that the wood grain is raised and the surface roughened.

Failing to follow manufacturers' product and application instructions when repainting exterior woodwork.

Using new colors that are inappropriate to the historic building or district.

Failing to undertake adequate measures to assure the protection of wood features.

Replacing an entire wood feature such as a cornice or wall when repair of the wood and limited replacement of deteriorated or missing parts are appropriate.

Using substitute material for the replacement part that does not convey the visual appearance of the surviving parts of the wood feature or that is physically or chemically incompatible.

Removing an entire wood feature that is unrepairable and not replacing it; or replacing it with a new feature that does not convey the same visual appearance.

Rehabilitation

The following work is highlighted to indicate that it represents the particularly complex technical or design aspects of **Rehabilitation** *projects and should only be considered after the preservation concerns listed above have been addressed.*

Recommended	*Not Recommended*
Design for the Replacement of Missing Historic Features	
Designing and installing a new wood feature such as a cornice or doorway when the historic feature is completely missing. It may be an accurate restoration using historical, pictorial, and physical documentation; or be a new design that is compatible with the size, scale, material, and color of the historic building.	Creating a false historical appearance because the replaced wood feature is based on insufficient historical, pictorial, and physical documentation. Introducing a new wood feature that is incompatible in size, scale, material and color.

Rehabilitation

Building Exterior

Architectural Metals: Cast iron, steel, pressed tin, copper, aluminum, and zinc

Recommended

Identifying, retaining, and preserving architectural metal features such as columns, capitals, window hoods, or stairways that are important in defining the overall historic character of the building; and their finishes and colors. Identification is also critical to differentiate between metals prior to work. Each metal has unique properties and thus requires different treatments.

Protecting and maintaining architectural metals from corrosion by providing proper drainage so that water does not stand on flat, horizontal surfaces or accumulate in curved, decorative features.

Cleaning architectural metals, when appropriate, to remove corrosion prior to repainting or applying other appropriate protective coatings.

Identifying the particular type of metal prior to any cleaning procedure and then testing to assure that the gentlest cleaning method possible is selected or determining that cleaning is inappropriate for the particular metal.

Not Recommended

Removing or radically changing architectural metal features which are important in defining the overall historic character of the building so that, as a result, the character is diminished.

Removing a major portion of the historic architectural metal from a facade instead of repairing or replacing only the deteriorated metal, then reconstructing the facade with new material in order to create a uniform, or "improved" appearance.

Radically changing the type of finish or its historic color or accent scheme.

Failing to identify, evaluate, and treat the causes of corrosion, such as moisture from leaking roofs or gutters.

Placing incompatible metals together without providing a reliable separation material. Such incompatibility can result in galvanic corrosion of the less noble metal, e.g., copper will corrode cast iron, steel, tin, and aluminum.

Exposing metals which were intended to be protected from the environment.

Applying paint or other coatings to metals such as copper, bronze, or stainless steel that were meant to be exposed.

Using cleaning methods which alter or damage the historic color, texture, and finish of the metal; or cleaning when it is inappropriate for the metal.

Removing the patina of historic metal. The patina may be a protective coating on some metals, such as bronze or copper, as well as a significant historic finish.

Rehabilitation

Recommended

Cleaning soft metals such as lead, tin, copper, terneplate, and zinc with appropriate chemical methods because their finishes can be easily abraded by blasting methods.

Using the gentlest cleaning methods for cast iron, wrought iron, and steel—hard metals—in order to remove paint buildup and corrosion. If handscraping and wire brushing have proven ineffective, low pressure grit blasting may be used as long as it does not abrade or damage the surface.

Applying appropriate paint or other coating systems after cleaning in order to decrease the corrosion rate of metals or alloys.

Repainting with colors that are appropriate to the historic building or district.

Applying an appropriate protective coating such as lacquer to an architectural metal feature such as a bronze door which is subject to heavy pedestrian use.

Evaluating the overall condition of the architectural metals to determine whether more than protection and maintenance are required, that is, if repairs to features will be necessary.

Repairing architectural metal features by patching, splicing, or otherwise reinforcing the metal following recognized preservation methods. Repairs may also include the limited replacement in kind—or with a compatible substitute material—of those extensively deteriorated or missing parts of features when there are surviving prototypes such as porch balusters, column capitals or bases; or porch cresting.

Not Recommended

Cleaning soft metals such as lead, tin, copper, terneplate, and zinc with grit blasting which will abrade the surface of the metal.

Failing to employ gentler methods prior to abrasively cleaning cast iron, wrought iron or steel; or using high pressure grit blasting.

Failing to re-apply protective coating systems to metals or alloys that require them after cleaning so that accelerated corrosion occurs.

Using new colors that are inappropriate to the historic building or district.

Failing to assess pedestrian use or new access patterns so that architectural metal features are subject to damage by use or inappropriate maintenance such as salting adjacent sidewalks.

Failing to undertake adequate measures to assure the protection of architectural metal features.

Replacing an entire architectural metal feature such as a column or a balustrade when repair of the metal and limited replacement of deteriorated or missing parts are appropriate.

Using a substitute material for the replacement part that does not convey the visual appearance of the surviving parts of the architectural metal feature or that is physically or chemically incompatible.

Rehabilitation

Recommended

Replacing in kind an entire architectural metal feature that is too deteriorated to repair—if the overall form and detailing are still evident—using the physical evidence as a model to reproduce the feature. Examples could include cast iron porch steps or steel sash windows. If using the same kind of material is not technically or economically feasible, then a compatible substitute material may be considered.

Not Recommended

Removing an architectural metal feature that is unrepairable and not replacing it; or replacing it with a new architectural metal feature that does not convey the same visual appearance.

The following work is highlighted to indicate that it represents the particularly complex technical or design aspects of **Rehabilitation** *projects and should only be considered after the preservation concerns listed above have been addressed.*

Recommended

Design for the Replacement of Missing Historic Features

Designing and installing a new architectural metal feature such as a metal cornice or cast iron capital when the historic feature is completely missing. It may be an accurate restoration using historical, pictorial, and physical documentation; or be a new design that is compatible with the size, scale, material, and color of the historic building.

Not Recommended

Creating a false historical appearance because the replaced architectural metal feature is based on insufficient historical, pictorial, and physical documentation.

Introducing a new architectural metal feature that is incompatible in size, scale, material, and color.

Rehabilitation

Building Exterior

Roofs

Recommended

Identifying, retaining, and preserving roofs—and their functional and decorative features—that are important in defining the overall historic character of the building. This includes the roof's shape, such as hipped, gambrel, and mansard; decorative features such as cupolas, cresting chimneys, and weathervanes; and roofing material such as slate, wood, clay tile, and metal, as well as its size, color, and patterning.

Protecting and maintaining a roof by cleaning the gutters and downspouts and replacing deteriorated flashing. Roof sheathing should also be checked for proper venting to prevent moisture condensation and water penetration; and to ensure that materials are free from insect infestation.

Providing adequate anchorage for roofing material to guard against wind damage and moisture penetration.

Protecting a leaking roof with plywood and building paper until it can be properly repaired.

Not Recommended

Radically changing, damaging, or destroying roofs which are important in defining the overall historic character of the building so that, as a result, the character is diminished.

Removing a major portion of the roof or roofing material that is repairable, then reconstructing it with new material in order to create a uniform, or "improved" appearance.

Changing the configuration of a roof by adding new features such as dormer windows, vents, or skylights so that the historic character is diminished.

Stripping the roof of sound historic material such as slate, clay tile, wood, and architectural metal.

Applying paint or other coatings to roofing material which has been historically uncoated.

Failing to clean and maintain gutters and downspouts properly so that water and debris collect and cause damage to roof fasteners, sheathing, and the underlying structure.

Allowing roof fasteners, such as nails and clips to corrode so that roofing material is subject to accelerated deterioration.

Permitting a leaking roof to remain unprotected so that accelerated deterioration of historic building materials—masonry, wood, plaster, paint and structural members—occurs.

Rehabilitation

Recommended

Repairing a roof by reinforcing the historic materials which comprise roof features. Repairs will also generally include the limited replacement in kind—or with compatible substitute material—of those extensively deteriorated or missing parts of features when there are surviving prototypes such as cupola louvers, dentils, dormer roofing; or slates, tiles, or wood shingles on a main roof.

Replacing in kind an entire feature of the roof that is too deteriorated to repair—if the overall form and detailing are still evident—using the physical evidence as a model to reproduce the feature. Examples can include a large section of roofing, or a dormer or chimney. If using the same kind of material is not technically or economically feasible, then a compatible substitute material may be considered.

Not Recommended

Replacing an entire roof feature such as a cupola or dormer when repair of the historic materials and limited replacement of deteriorated or missing parts are appropriate.

Failing to reuse intact slate or tile when only the roofing substrate needs replacement.

Using a substitute material for the replacement part that does not convey the visual appearance of the surviving parts of the roof or that is physically or chemically incompatible.

Removing a feature of the roof that is unrepairable, such as a chimney or dormer, and not replacing it; or replacing it with a new feature that does not convey the same visual appearance.

Rehabilitation

The following work is highlighted to indicate that it represents the particularly complex technical or design aspects of Rehabilitation projects and should only be considered after the preservation concerns listed above have been addressed.

Recommended	*Not Recommended*
Design for the Replacement of Missing Historic Features	
Designing and constructing a new feature when the historic feature is completely missing, such as chimney or cupola. It may be an accurate restoration using historical, pictorial, and physical documentation; or be a new design that is compatible with the size, scale, material, and color of the historic building.	Creating a false historical appearance because the replaced feature is based on insufficient historical, pictorial, and physical documentation.
	Introducing a new roof feature that is incompatible in size, scale, material and color.
Alterations/Additions for the New Use	
Installing mechanical and service equipment on the roof such as air conditioning, transformers, or solar collectors when required for the new use so that they are inconspicuous from the public right-of-way and do not damage or obscure character-defining features.	Installing mechanical or service equipment so that it damages or obscures character-defining features; or is conspicuous from the public right-of-way.
Designing additions to roofs such as residential, office, or storage spaces; elevator housing; decks and terraces; or dormers or skylights when required by the new use so that they are inconspicuous from the public right-of-way and do not damage or obscure character-defining features.	Radically changing a character-defining roof shape or damaging or destroying character-defining roofing material as a result of incompatible design or improper installation techniques.

Rehabilitation

Building Exterior

Windows

Recommended

Identifying, retaining, and preserving windows—and their functional and decorative features—that are important in defining the overall historic character of the building. Such features can include frames, sash, muntins, glazing, sills, heads, hoodmolds, panelled or decorated jambs and moldings, and interior and exterior shutters and blinds.

Conducting an indepth survey of the condition of existing windows early in rehabilitation planning so that repair and upgrading methods and possible replacement options can be fully explored.

Protecting and maintaining the wood and architectural metals which comprise the window frame, sash, muntins, and surrounds through appropriate surface treatments such as cleaning, rust removal, limited paint removal, and re-application of protective coating systems.

Making windows weathertight by re-caulking and replacing or installing weatherstripping. These actions also improve thermal efficiency.

Not Recommended

Removing or radically changing windows which are important in defining the historic character of the building so that, as a result, the character is diminished.

Changing the number, location, size or glazing pattern of windows, through cutting new openings, blocking-in windows, and installing replacement sash that do not fit the historic window opening.

Changing the historic appearance of windows through the use of inappropriate designs, materials, finishes, or colors which noticeably change the sash, depth of reveal, and muntin configuration; the reflectivity and color of the glazing; or the appearance of the frame.

Obscuring historic window trim with metal or other material.

Stripping windows of historic material such as wood, cast iron, and bronze.

Replacing windows solely because of peeling paint, broken glass, stuck sash, and high air infiltration. These conditions, in themselves, are no indication that windows are beyond repair.

Failing to provide adequate protection of materials on a cyclical basis so that deterioration of the window results.

Retrofitting or replacing windows rather than maintaining the sash, frame, and glazing.

Rehabilitation

Recommended

Evaluating the overall condition of materials to determine whether more than protection and maintenance are required, i.e. if repairs to windows and window features will be required.

Repairing window frames and sash by patching, splicing, consolidating or otherwise reinforcing. Such repair may also include replacement in kind—or with compatible substitute material—of those parts that are either extensively deteriorated or are missing when there are surviving prototypes such as architraves, hoodmolds, sash, sills, and interior or exterior shutters and blinds.

Replacing in kind an entire window that is too deteriorated to repair using the same sash and pane configuration and other design details. If using the same kind of material is not technically or economically feasible when replacing windows deteriorated beyond repair, then a compatible substitute material may be considered.

Not Recommended

Failing to undertake adequate measures to assure the protection of historic windows.

Replacing an entire window when repair of materials and limited replacement of deteriorated or missing parts are appropriate.

Failing to reuse serviceable window hardware such as brass sash lifts and sash locks.

Using substitute material for the replacement part that does not convey the visual appearance of the surviving parts of the window or that is physically or chemically incompatible.

Removing a character-defining window that is unrepairable and blocking it in; or replacing it with a new window that does not convey the same visual appearance.

Rehabilitation

The following work is highlighted to indicate that it represents the particularly complex technical or design aspects of **Rehabilitation** *projects and should only be considered after the preservation concerns listed above have been addressed.*

Recommended	*Not Recommended*
Design for the Replacement of Missing Historic Features	
Designing and installing new windows when the historic windows (frames, sash and glazing) are completely missing. The replacement windows may be an accurate restoration using historical, pictorial, and physical documentation; or be a new design that is compatible with the window openings and the historic character of the building.	Creating a false historical appearance because the replaced window is based on insufficient historical, pictorial, and physical documentation. Introducing a new design that is incompatible with the historic character of the building.
Alterations/Additions for the New Use	
Designing and installing additional windows on rear or other non-character-defining elevations if required by the new use. New window openings may also be cut into exposed party walls. Such design should be compatible with the overall design of the building, but not duplicate the fenestration pattern and detailing of a character-defining elevation.	Installing new windows, including frames, sash, and muntin configuration that are incompatible with the building's historic appearance or obscure, damage, or destroy character-defining features.
Providing a setback in the design of dropped ceilings when they are required for the new use to allow for the full height of the window openings.	Inserting new floors or furred-down ceilings which cut across the glazed areas of windows so that the exterior form and appearance of the windows are changed.

Building Exterior *Windows* 83

Rehabilitation

(a) An armory complex was rehabilitated for rental housing. (b) This view of the rear elevation shows the paired, nine-over-nine wood sash windows and high sills that characterized the building. (c) After inappropriate rehabilitation work, the same rear elevation is shown with new skylights added to the roof, prefabricated panels filling the former brick areas, and new wood decks and privacy fences. Because the work changed the historic character, the project did not meet the Standards.

Rehabilitation

Building Exterior

Entrances and Porches

Recommended

Identifying, retaining, and preserving entrances and porches—and their functional and decorative features—that are important in defining the overall historic character of the building such as doors, fanlights, sidelights, pilaster, entablatures, columns, balustrades, and stairs.

Protecting and maintaining the masonry, wood, and architectural metals that comprise entrances and porches through appropriate surface treatments such as cleaning, rust removal, limited paint removal, and re-application of protective coating systems.

Evaluating the overall condition of materials to determine whether more than protection and maintenance are required, that is, repairs to entrance and porch features will be necessary.

Repairing entrances and porches by reinforcing the historic materials. Repair will also generally include the limited replacement in kind—or with compatible substitute material—of those extensively deteriorated or missing parts of repeated features where there are surviving prototypes such as balustrades, cornices, entablatures, columns, sidelights, and stairs.

Not Recommended

Removing or radically changing entrances and porches which are important in defining the overall historic character of the building so that, as a result, the character is diminished.

Stripping entrances and porches of historic material such as wood, cast iron, terra cotta tile, and brick.

Removing an entrance or porch because the building has been re-oriented to accommodate a new use.

Cutting new entrances on a primary elevation.

Altering utilitarian or service entrances so they appear to be formal entrances by adding panelled doors, fanlights, and sidelights.

Failing to provide adequate protection to materials on a cyclical basis so that deterioration of entrances and porches results.

Failing to undertake adequate measures to assure the protection of historic entrances and porches.

Replacing an entire entrance or porch when the repair of materials and limited replacement of parts are appropriate.

Using a substitute material for the replacement parts that does not convey the visual appearance of the surviving parts of the entrance and porch or that is physically or chemically incompatible.

Rehabilitation

In Rehabilitation, deteriorated features should be repaired, whenever possible, and replaced when the severity of the damage makes it necessary. Here, a two-story porch is seen prior to treatment (left). The floor boards are rotted out and the columns are in a state of collapse, supported only by crude, temporary shafts. Other components are in varying stages of decay. Appropriate work on the historic porch (right) included repairs to the porch rails; and total replacement of the extensively deteriorated columns and floor boards. Some dismantling of the porch was necessary.

Rehabilitation

Recommended

Replacing in kind an entire entrance or porch that is too deteriorated to repair—if the form and detailing are still evident—using the physical evidence as a model to reproduce the feature. If using the same kind of material is not technically or economically feasible, then a compatible substitute material may be considered.

Not Recommended

Removing an entrance or porch that is unrepairable and not replacing it; or replacing it with a new entrance or porch that does not convey the same visual appearance.

The following work is highlighted to indicate that it represents the particularly complex technical or design aspects of **Rehabilitation** *projects and should only be considered after the preservation concerns listed above have been addressed.*

Recommended

Design for the Replacement of Missing Historic Features

Designing and constructing a new entrance or porch when the historic entrance or porch is completely missing. It may be a restoration based on historical, pictorial, and physical documentation; or be a new design that is compatible with the historic character building.

Alterations/Additions for the New Use

Designing enclosures for historic porches on secondary elevations when required by the new use in a manner that preserves the historic character of the building. This can include using large sheets of glass and recessing the enclosure wall behind existing scrollwork, posts, and balustrades.

Designing and installing additional entrances or porches on secondary elevations when required for the new use in a manner that preserves the historic character of the buildings, i.e., limiting such alteration to non-character-defining elevations.

Not Recommended

Creating a false historical appearance because the replaced entrance or porch is based on insufficient historical, pictorial, and physical documentation.

Introducing a new entrance or porch that is incompatible in size, scale, material, and color.

Enclosing porches in a manner that results in a diminution or loss of historic character by using materials such as wood, stucco, or masonry.

Installing secondary service entrances and porches that are incompatible in size and scale with the historic building or obscure, damage, or destroy character-defining features.

Rehabilitation

Building Exterior

Storefronts

Recommended

Identifying, retaining, and preserving storefronts—and their functional and decorative features—that are important in defining the overall historic character of the building such as display windows, signs, doors, transoms, kick plates, corner posts, and entablatures. The removal of inappropriate, non-historic cladding, false mansard roofs, and other later alterations can help reveal the historic character of a storefront.

Protecting and maintaining masonry, wood, and architectural metals which comprise storefronts through appropriate treatments such as cleaning, rust removal, limited paint removal, and reapplication of protective coating systems.

Protecting storefronts against arson and vandalism before work begins by boarding up windows and installing alarm systems that are keyed into local protection agencies.

Evaluating the existing condition of storefront materials to determine whether more than protection and maintenance are required, that is, if repairs to features will be necessary.

Not Recommended

Removing or radically changing storefronts—and their features—which are important in defining the overall historic character of the building so that, as a result, the character is diminished.

Changing the storefront so that it appears residential rather than commercial in character.

Removing historic material from the storefront to create a recessed arcade.

Introducing coach lanterns, mansard designs, wood shakes, nonoperable shutters, and small-paned windows if they cannot be documented historically.

Changing the location of a storefront's main entrance.

Failing to provide adequate protection of materials on a cyclical basis so that deterioration of storefront features results.

Permitting entry into the building through unsecured or broken windows and doors so that interior features and finishes are damaged by exposure to weather or vandalism.

Stripping storefronts of historic material such as wood, cast iron, terra cotta, carrara glass, and brick.

Failing to undertake adequate measures to assure the preservation of the historic storefront.

Rehabilitation

Recommended

Repairing storefronts by reinforcing the historic materials. Repairs will also generally include the limited replacement in kind—or with compatible substitute materials—of those extensively deteriorated or missing parts of storefronts where there are surviving prototypes such as transoms, kick plates, pilasters, or signs.

Replacing in kind an entire storefront that is too deteriorated to repair—if the overall form and detailing are still evident—using the physical evidence as a model. If using the same material is not technically or economically feasible, then compatible substitute materials may be considered.

Not Recommended

Replacing an entire storefront when repair of materials and limited replacement of its parts are appropriate.

Using substitute material for the replacement parts that does not convey the same visual appearance as the surviving parts of the storefront or that is physically or chemically incompatible.

Removing a storefront that is unrepairable and not replacing it; or replacing it with a new storefront that does not convey the same visual appearance.

The following work is highlighted to indicate that it represents the particularly complex technical or design aspects of **Rehabilitation** *projects and should only be considered after the preservation concerns listed above have been addressed.*

Recommended

Design for the Replacement of Missing Historic Features

Designing and constructing a new storefront when the historic storefront is completely missing. It may be an accurate restoration using historical, pictorial, and physical documentation; or be a new design that is compatible with the size, scale, material, and color of the historic building.

Not Recommended

Creating a false historical appearance because the replaced storefront is based on insufficient historical, pictorial, and physical documentation.

Introducing a new design that is incompatible in size, scale, material, and color.

Using inappropriately scaled signs and logos or other types of signs that obscure, damage, or destroy remaining character-defining features of the historic building.

Rehabilitation

a

b

c

In the treatment, Rehabilitation, one option for replacing missing historic features is to use pictorial documentation and/or physical evidence to re-create the historic feature. (a) In this example, the ornamental cornice of an 1866 limestone building was missing; and the ground level storefront had been extensively altered. (b) and (c) Based on the availability of photographic and other documentation, the owners were able to accurately restore the cornice and storefront to their historic configuration. A substitute material, fiberglass, was used to fabricate the missing pressed metal cornice, an acceptable alternative in this project. All work met the Standards.

Rehabilitation

Building Interior
Structural Systems

Recommended

Identifying, retaining, and preserving structural systems—and individual features of systems—that are important in defining the overall historic character of the building, such as post and beam systems, trusses, summer beams, vigas, cast iron columns, above-grade stone foundation walls, or loadbearing brick or stone walls.

Protecting and maintaining the structural system by cleaning the roof gutters and downspouts; replacing roof flashing; keeping masonry, wood, and architectural metals in a sound condition; and ensuring that structural members are free from insect infestation.

Examining and evaluating the physical condition of the structural system and its individual features using non-destructive techniques such as X-ray photography.

Not Recommended

Removing, covering, or radically changing visible features of structural systems which are important in defining the overall historic character of the building so that, as a result, the character is diminished.

Putting a new use into the building which could overload the existing structural system; or installing equipment or mechanical systems which could damage the structure.

Demolishing a loadbearing masonry wall that could be augmented and retained, and replacing it with a new wall (i.e., brick or stone), using the historic masonry only as an exterior veneer.

Leaving known structural problems untreated such as deflection of beams, cracking and bowing of walls, or racking of structural members.

Utilizing treatments or products that accelerate the deterioration of structural material such as introducing urea-formaldehyde foam insulation into frame walls.

Failing to provide proper building maintenance so that deterioration of the structural system results. Causes of deterioration include subsurface ground movement, vegetation growing too close to foundation walls, improper grading, fungal rot, and poor interior ventilation that results in condensation.

Utilizing destructive probing techniques that will damage or destroy structural material.

Rehabilitation

Recommended

Repairing the structural system by augmenting or upgrading individual parts or features. For example, weakened structural members such as floor framing can be paired with a new member, braced, or otherwise supplemented and reinforced.

Replacing in kind—or with substitute material—those portions or features of the structural system that are either extensively deteriorated or are missing when there are surviving prototypes such as cast iron columns, roof rafters or trusses, or sections of loadbearing walls. Substitute material should convey the same form, design, and overall visual appearance as the historic feature; and, at a minimum, be equal to its loadbearing capabilities.

Not Recommended

Upgrading the building structurally in a manner that diminishes the historic character of the exterior, such as installing strapping channels or removing a decorative cornice; or damages interior features or spaces.

Replacing a structural member or other feature of the structural system when it could be augmented and retained.

Installing a visible replacement feature that does not convey the same visual appearance, e.g., replacing an exposed wood summer beam with a steel beam.

Using substitute material that does not equal the loadbearing capabilities of the historic material and design or is otherwise physically or chemically incompatible.

Rehabilitation

The following work is highlighted to indicate that it represents the particularly complex technical or design aspects of **Rehabilitation** *projects and should only be considered after the preservation concerns listed above have been addressed.*

Recommended	*Not Recommended*
Alterations/Additions for the New Use	
Limiting any new excavations adjacent to historic foundations to avoid undermining the structural stability of the building or adjacent historic buildings. Studies should be done to ascertain potential damage to archeological resources.	Carrying out excavations or regrading adjacent to or within a historic building which could cause the historic foundation to settle, shift, or fail; could have a similar effect on adjacent historic buildings; or could destroy significant archeological resources.
Correcting structural deficiencies in preparation for the new use in a manner that preserves the structural system and individual character-defining features.	Radically changing interior spaces or damaging or destroying features or finishes that are character-defining while trying to correct structural deficiencies in preparation for the new use.
Designing and installing new mechanical or electrical systems when required for the new use which minimize the number of cutouts or holes in structural members.	Installing new mechanical and electrical systems or equipment in a manner which results in numerous cuts, splices, or alterations to the structural members.
Adding a new floor when required for the new use if such an alteration does not damage or destroy the structural system or obscure, damage, or destroy character-defining spaces, features, or finishes.	Inserting a new floor when such a radical change damages a structural system or obscures or destroys interior spaces, features, or finishes.
	Inserting new floors or furred-down ceilings which cut across the glazed areas of windows so that the exterior form and appearance of the windows are radically changed.
Creating an atrium or a light well to provide natural light when required for the new use in a manner that assures the preservation of the structural system as well as character-defining interior spaces, features, and finishes.	Damaging the structural system or individual features; or radically changing, damaging, or destroying character-defining interior spaces, features, or finishes in order to create an atrium or a light well.

Building Interior *Structural Systems*

Rehabilitation

Building Interior

Spaces, Features, and Finishes

Recommended

Interior Spaces

Identifying, retaining, and preserving a floor plan or interior spaces that are important in defining the overall historic character of the building. This includes the size, configuration, proportion, and relationship of rooms and corridors; the relationship of features to spaces; and the spaces themselves such as lobbies, reception halls, entrance halls, double parlors, theaters, auditoriums, and important industrial or commercial spaces.

Interior Features and Finishes

Identifying, retaining, and preserving interior features and finishes that are important in defining the overall historic character of the building, including columns, cornices, baseboards, fireplaces and mantels, panelling, light fixtures, hardware, and flooring; and wallpaper, plaster, paint, and finishes such as stencilling, marbling, and graining; and other decorative materials that accent interior features and provide color, texture, and patterning to walls, floors, and ceilings.

Not Recommended

Radically changing a floor plan or interior spaces—including individual rooms—which are important in defining the overall historic character of the building so that, as a result, the character is diminished.

Altering the floor plan by demolishing principal walls and partitions to create a new appearance.

Altering or destroying interior spaces by inserting floors, cutting through floors, lowering ceilings, or adding or removing walls.

Relocating an interior feature such as a staircase so that the historic relationship between features and spaces is altered.

Removing or radically changing features and finishes which are important in defining the overall historic character of the building so that, as a result, the character is diminished.

Installing new decorative material that obscures or damages character-defining interior features or finishes.

Removing paint, plaster, or other finishes from historically finished surfaces to create a new appearance (e.g., removing plaster to expose masonry surfaces such as brick walls or a chimney piece).

Applying paint, plaster, or other finishes to surfaces that have been historically unfinished to create a new appearance.

Stripping paint to bare wood rather than repairing or reapplying grained or marbled finishes to features such as doors and panelling.

Radically changing the type of finish or its color, such as painting a previously varnished wood feature.

Rehabilitation

Recommended

Protecting and maintaining masonry, wood, and architectural metals which comprise interior features through appropriate surface treatments such as cleaning, rust removal, limited paint removal, and reapplication of protective coating systems.

Protecting interior features and finishes against arson and vandalism before project work begins, erecting protective fencing, boarding-up windows, and installing fire alarm systems that are keyed to local protection agencies.

Protecting interior features such as a staircase, mantel, or decorative finishes and wall coverings against damage during project work by covering them with heavy canvas or plastic sheets.

Not Recommended

Failing to provide adequate protection to materials on a cyclical basis so that deterioration of interior features results.

Permitting entry into historic buildings through unsecured or broken windows and doors so that the interior features and finishes are damaged by exposure to weather or vandalism.

Stripping interiors of features such as woodwork, doors, windows, light fixtures, copper piping, radiators; or of decorative materials.

Failing to provide proper protection of interior features and finishes during work so that they are gouged, scratched, dented, or otherwise damaged.

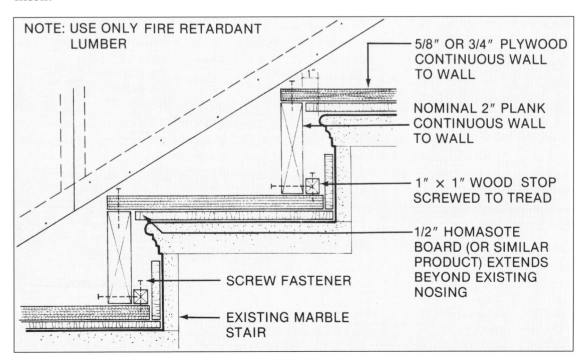

Historic features that characterize a building should always be protected from damage during rehabilitation work. The drawing shows how a resilient, temporary stair covering was applied over the existing marble staircase. Drawing: National Park Service staff, based on material originally prepared by Emery Roth and Sons, P.C.

Building Interior *Spaces, Features, and Finishes*

Rehabilitation

Recommended	*Not Recommended*
Installing protective coverings in areas of heavy pedestrian traffic to protect historic features such as wall coverings, parquet flooring and panelling.	Failing to take new use patterns into consideration so that interior features and finishes are damaged.
Removing damaged or deteriorated paints and finishes to the next sound layer using the gentlest method possible, then repainting or refinishing using compatible paint or other coating systems.	Using destructive methods such as propane or butane torches or sandblasting to remove paint or other coatings. These methods can irreversibly damage the historic materials that comprise interior features.
Repainting with colors that are appropriate to the historic building.	Using new paint colors that are inappropriate to the historic building.
Limiting abrasive cleaning methods to certain industrial warehouse buildings where the interior masonry or plaster features do not have distinguishing design, detailing, tooling, or finishes; and where wood features are not finished, molded, beaded, or worked by hand. Abrasive cleaning should only be considered after other, gentler methods have been proven ineffective.	Changing the texture and patina of character-defining features through sandblasting or use of abrasive methods to remove paint, discoloration or plaster. This includes both exposed wood (including structural members) and masonry.
Evaluating the existing condition of materials to determine whether more than protection and maintenance are required, that is, if repairs to interior features and finishes will be necessary.	Failing to undertake adequate measures to assure the protection of interior features and finishes.
Repairing interior features and finishes by reinforcing the historic materials. Repair will also generally include the limited replacement in kind—or with compatible substitute material—of those extensively deteriorated or missing parts of repeated features when there are surviving prototypes such as stairs, balustrades, wood panelling, columns; or decorative wall coverings or ornamental tin or plaster ceilings.	Replacing an entire interior feature such as a staircase, panelled wall, parquet floor, or cornice; or finish such as a decorative wall covering or ceiling when repair of materials and limited replacement of such parts are appropriate. Using a substitute material for the replacement part that does not convey the visual appearance of the surviving parts or portions of the interior feature or finish or that is physically or chemically incompatible.

Rehabilitation

Recommended

Replacing in kind an entire interior feature or finish that is too deteriorated to repair—if the overall form and detailing are still evident—using the physical evidence as a model for reproduction. Examples could include wainscoting, a tin ceiling, or interior stairs. If using the same kind of material is not technically or economically feasible, then a compatible substitute material may be considered.

Not Recommended

Removing a character-defining feature or finish that is unrepairable and not replacing it; or replacing it with a new feature or finish that does not convey the same visual appearance.

a

b

Rehabilitating historic dwelling units often includes some level of lead-paint hazard abatement. Whenever lead-base paint begins to peel, chip, craze, or otherwise comes loose (a), it should be removed in a manner that protects the worker as well as the immediate environment. In this example (b), the deteriorating lead-paint was removed throughout the apartment building and a compatible primer and finish paint applied. Photos: Sharon C. Park, AIA.

Building Interior *Spaces, Features, and Finishes*

Rehabilitation

The following work is highlighted to indicate that it represents the particularly complex technical or design aspects of **Rehabilitation** *projects and should only be considered after the preservation concerns listed above have been addressed.*

Recommended	*Not Recommended*
Design for the Replacement of Missing Historic Features	
Designing and installing a new interior feature or finish if the historic feature or finish is completely missing. This could include missing partitions, stairs, elevators, lighting fixtures, and wall coverings; or even entire rooms if all historic spaces, features, and finishes are missing or have been destroyed by inappropriate "renovations." The design may be a restoration based on historical, pictorial, and physical documentation; or be a new design that is compatible with the historic character of the building, district, or neighborhood.	Creating a false historical appearance because the replaced feature is based on insufficient physical, historical, and pictorial documentation or on information derived from another building.
	Introducing a new interior feature or finish that is incompatible with the scale, design, materials, color, and texture of the surviving interior features and finishes.
Alterations/Additions for the New Use	
Accommodating service functions such as bathrooms, mechanical equipment, and office machines required by the building's new use in secondary spaces such as first floor service areas or on upper floors.	Dividing rooms, lowering ceilings, and damaging or obscuring character-defining features such as fireplaces, niches, stairways or alcoves, so that a new use can be accommodated in the building.
Reusing decorative material or features that have had to be removed during the rehabilitation work including wall and baseboard trim, door molding, panelled doors, and simple wainscoting; and relocating such material or features in areas appropriate to their historic placement.	Discarding historic material when it can be reused within the rehabilitation project or relocating it in historically inappropriate areas.
Installing permanent partitions in secondary spaces; removable partitions that do not destroy the sense of space should be installed when the new use requires the subdivision of character-defining interior space.	Installing permanent partitions that damage or obscure character-defining spaces, features, or finishes.
Enclosing an interior stairway where required by code so that its character is retained. In many cases, glazed fire-rated walls may be used.	Enclosing an interior stairway with fire-rated construction so that the stairwell space or any character-defining features are destroyed.

Rehabilitation

Recommended	Not Recommended
Placing new code-required stairways or elevators in secondary and service areas of the historic building.	Radically changing, damaging, or destroying character-defining spaces, features, or finishes when adding new code-required stairways and elevators.
Creating an atrium or a light well to provide natural light when required for the new use in a manner that preserves character-defining interior spaces, features, and finishes as well as the structural system.	Destroying character-defining interior spaces, features, or finishes; or damaging the structural system in order to create an atrium or light well.
Adding a new floor if required for the new use in a manner that preserves character-defining structural features, and interior spaces, features, and finishes.	Inserting a new floor within a building that alters or destroys the fenestration; radically changes a character-defining interior space; or obscures, damages, or destroys decorative detailing.

Rehabilitation

Building Interior

Mechanical Systems: Heating, Air Conditioning, Electrical, and Plumbing

Recommended	*Not Recommended*
Identifying, retaining, and preserving visible features of early mechanical systems that are important in defining the overall historic character of the building, such as radiators, vents, fans, grilles, plumbing fixtures, switchplates, and lights.	Removing or radically changing features of mechanical systems that are important in defining the overall historic character of the building so that, as a result, the character is diminished.
Protecting and maintaining mechanical, plumbing, and electrical systems and their features through cyclical cleaning and other appropriate measures.	Failing to provide adequate protection of materials on a cyclical basis so that deterioration of mechanical systems and their visible features results.
Preventing accelerated deterioration of mechanical systems by providing adequate ventilation of attics, crawlspaces, and cellars so that moisture problems are avoided.	Enclosing mechanical systems in areas that are not adequately ventilated so that deterioration of the systems results.
Improving the energy efficiency of existing mechanical systems to help reduce the need for elaborate new equipment. Consideration should be given to installing storm windows, insulating attic crawl space, or adding awnings, if appropriate.	Installing unnecessary air conditioning or climate control systems which can add excessive moisture to the building. This additional moisture can either condense inside, damaging interior surfaces, or pass through interior walls to the exterior, potentially damaging adjacent materials as it migrates.
Repairing mechanical systems by augmenting or upgrading system parts, such as installing new pipes and ducts; rewiring; or adding new compressors or boilers.	Replacing a mechanical system or its functional parts when it could be upgraded and retained.
Replacing in kind—or with compatible substitute material—those visible features of mechanical systems that are either extensively deteriorated or are prototypes such as ceiling fans, switchplates, radiators, grilles, or plumbing fixtures.	Installing a visible replacement feature that does not convey the same visual appearance.

Rehabilitation

The following work is highlighted to indicate that it represents the particularly complex technical or design aspects of **Rehabilitation** *projects and should only be considered after the preservation concerns listed above have been addressed.*

Recommended	*Not Recommended*
Alterations/Additions for the New Use	
Installing a completely new mechanical system if required for the new use so that it causes the least alteration possible to the building's floor plan, the exterior elevations, and the least damage to the historic building material.	Installing a new mechanical system so that character-defining structural or interior features are radically changed, damaged, or destroyed.
Providing adequate structural support for new mechanical equipment.	Failing to consider the weight and design of new mechanical equipment so that, as a result, historic structural members or finished surfaces are weakened or cracked.
Installing the vertical runs of ducts, pipes, and cables in closets, service rooms, and wall cavities.	Installing vertical runs of ducts, pipes, and cables in places where they will obscure character-defining features.
	Concealing mechanical equipment in walls or ceilings in a manner that requires the removal of historic building material.
	Installing a "dropped" acoustical ceiling to hide mechanical equipment when this destroys the proportions of character-defining interior spaces.
Installing air conditioning units if required by the new use in such a manner that historic features are not damaged or obscured and excessive moisture is not generated that will accelerate deterioration of historic materials.	Cutting through features such as masonry walls in order to install air conditioning units.
Installing heating/air conditioning units in the window frames in such a manner that the sash and frames are protected. Window installations should be considered only when all other viable heating/cooling systems would result in significant damage to historic materials.	Radically changing the appearance of the historic building or damaging or destroying windows by installing heating/air conditioning units in historic window frames.

Building Interior *Mechanical Systems*

Rehabilitation

Building Site

Recommended

Identifying, retaining, and preserving buildings and their features as well as features of the site that are important in defining its overall historic character. Site features may include circulation systems such as walks, paths, roads, or parking; vegetation such as trees, shrubs, fields, or herbaceous plant material; landforms such as terracing, berms or grading; furnishings such as lights, fences, or benches; decorative elements such as sculpture, statuary or monuments; water features including fountains, streams, pools, or lakes; and subsurface archeological features which are important in defining the history of the site.

Retaining the historic relationship between buildings and the landscape.

Protecting and maintaining buildings and the site by providing proper drainage to assure that water does not erode foundation walls; drain toward the building; or damage or erode the landscape.

Minimizing disturbance of terrain around buildings or elsewhere on the site, thus reducing the possibility of destroying or damaging important landscape features or archeological resources.

Not Recommended

Removing or radically changing buildings and their features or site features which are important in defining the overall historic character of the property so that, as a result, the character is diminished.

Removing or relocating buildings or landscape features, thus destroying the historic relationship between buildings and the landscape.

Removing or relocating historic buildings on a site or in a complex of related historic structures—such as a mill complex or farm—thus diminishing the historic character of the site or complex.

Moving buildings onto the site, thus creating a false historical appearance.

Radically changing the grade level of the site. For example, changing the grade adjacent to a building to permit development of a formerly below-grade area that would drastically change the historic relationship of the building to its site.

Failing to maintain adequate site drainage so that buildings and site features are damaged or destroyed; or alternatively, changing the site grading so that water no longer drains properly.

Introducing heavy machinery into areas where it may disturb or damage important landscape features or archeological resources.

Rehabilitation

Recommended

Surveying and documenting areas where the terrain will be altered to determine the potential impact to important landscape features or archeological resources.

Protecting, e.g., preserving in place important archeological resources.

Planning and carrying out any necessary investigation using professional archeologists and modern archeological methods when preservation in place is not feasible.

Preserving important landscape features, including ongoing maintenance of historic plant material.

Protecting the building and landscape features against arson and vandalism before rehabilitation work begins, i.e., erecting protective fencing and installing alarm systems that are keyed into local protection agencies.

Providing continued protection of historic building materials and plant features through appropriate cleaning, rust removal, limited paint removal, and re-application of protective coating systems; and pruning and vegetation management.

Evaluating the overall condition of the materials and features of the property to determine whether more than protection and maintenance are required, that is, if repairs to building and site features will be necessary.

Not Recommended

Failing to survey the building site prior to the beginning of rehabilitation work which results in damage to, or destruction of, important landscape features or archeological resources.

Leaving known archeological material unprotected so that it is damaged during rehabilitation work.

Permitting unqualified personnel to perform data recovery on archeological resources so that improper methodology results in the loss of important archeological material.

Allowing important landscape features to be lost or damaged due to a lack of maintenance.

Permitting the property to remain unprotected so that the building and landscape features or archeological resources are damaged or destroyed.

Removing or destroying features from the building or site such as wood siding, iron fencing, masonry balustrades, or plant material.

Failing to provide adequate protection of materials on a cyclical basis so that deterioration of building and site features results.

Failing to undertake adequate measures to assure the protection of building and site features.

Rehabilitation

Recommended

Repairing features of the building and site by reinforcing historic materials.

Replacing in kind an entire feature of the building or site that is too deteriorated to repair if the overall form and detailing are still evident. Physical evidence from the deteriorated feature should be used as a model to guide the new work. This could include an entrance or porch, walkway, or fountain. If using the same kind of material is not technically or economically feasible, then a compatible substitute material may be considered.

Replacing deteriorated or damaged landscape features in kind.

Not Recommended

Replacing an entire feature of the building or site such as a fence, walkway, or driveway when repair of materials and limited compatible replacement of deteriorated or missing parts are appropriate.

Using a substitute material for the replacement part that does not convey the visual appearance of the surviving parts of the building or site feature or that is physically or chemically incompatible.

Removing a feature of the building or site that is unrepairable and not replacing it; or replacing it with a new feature that does not convey the same visual appearance.

Adding conjectural landscape features to the site such as period reproduction lamps, fences, fountains, or vegetation that are historically inappropriate, thus creating a false sense of historic development.

Rehabilitation

The following work is highlighted to indicate that it represents the particularly complex technical or design aspects of **Rehabilitation** *project work and should only be considered after the preservation concerns listed above have been addressed.*

Recommended	*Not Recommended*
Design for the Replacement of Missing Historic Features	
Designing and constructing a new feature of a building or site when the historic feature is completely missing, such as an outbuilding, terrace, or driveway. It may be based on historical, pictorial, and physical documentation; or be a new design that is compatible with the historic character of the building and site.	Creating a false historical appearance because the replaced feature is based on insufficient historical, pictorial, and physical documentation. Introducing a new building or site feature that is out of scale or of an otherwise inappropriate design. Introducing a new landscape feature, including plant material, that is visually incompatible with the site, or that alters or destroys the historic site patterns or vistas.
Alterations/Additions for the New Use	
Designing new onsite parking, loading docks, or ramps when required by the new use so that they are as unobtrusive as possible and assure the preservation of the historic relationship between the building or buildings and the landscape. Designing new exterior additions to historic buildings or adjacent new construction which is compatible with the historic character of the site and which preserves the historic relationship between the building or buildings and the landscape.	Locating any new construction on the building site in a location which contains important landscape features or open space, for example removing a lawn and walkway and installing a parking lot. Placing parking facilities directly adjacent to historic buildings where automobiles may cause damage to the buildings or landscape features, or be intrusive to the building site. Introducing new construction onto the building site which is visually incompatible in terms of size, scale, design, materials, color, and texture; which destroys historic relationships on the site; or which damages or destroys important landscape features.
Removing non-significant buildings, additions, or site features which detract from the historic character of the site.	Removing a historic building in a complex of buildings; or removing a building feature, or a landscape feature which is important in defining the historic character of the site.

Rehabilitation

Setting (District/Neighborhood)

Recommended

Identifying retaining, and preserving building and landscape features which are important in defining the historic character of the setting. Such features can include roads and streets, furnishings such as lights or benches, vegetation, gardens and yards, adjacent open space such as fields, parks, commons or woodlands, and important views or visual relationships.

Retaining the historic relationship between buildings and landscape features of the setting. For example, preserving the relationship between a town common and its adjacent historic houses, municipal buildings, historic roads, and landscape features.

Protecting and maintaining historic building materials and plant features through appropriate cleaning, rust removal, limited paint removal, and reapplication of protective coating systems; and pruning and vegetation management.

Protecting building and landscape features such as lighting or trees, against arson and vandalism before rehabilitation work begins by erecting protective fencing and installing alarm systems that are keyed into local protection agencies.

Evaluating the overall condition of the building and landscape features to determine whether more than protection and maintenance are required, that is, if repairs to features will be necessary.

Not Recommended

Removing or radically changing those features of the setting which are important in defining the historic character.

Destroying the relationship between the buildings and landscape features within the setting by widening existing streets, changing landscape materials or constructing inappropriately located new streets or parking.

Removing or relocating historic buildings or landscape features, thus destroying their historic relationship within the setting.

Failing to provide adequate protection of materials on a cyclical basis which results in the deterioration of building and landscape features.

Permitting the building and setting to remain unprotected so that interior or exterior features are damaged.

Stripping or removing features from buildings or the setting such as wood siding, iron fencing, terra cotta balusters, or plant material.

Failing to undertake adequate measures to assure the protection of building and landscape features.

Rehabilitation

Recommended

Repairing features of the building and landscape by reinforcing the historic materials. Repair will also generally include the replacement in kind—or with a compatible substitute material—of those extensively deteriorated or missing parts of features when there are surviving prototypes such as porch balustrades or paving materials.

Replacing in kind an entire feature of the building or landscape that is too deteriorated to repair— when the overall form and detailing are still evident —using the physical evidence as a model to guide the new work. If using the same kind of material is not technically or economically feasible, then a compatible substitute material may be considered.

Not Recommended

Replacing an entire feature of the building or landscape when repair of materials and limited replacement of deteriorated or missing parts are appropriate.

Using a substitute material for the replacement part that does not convey the visual appearance of the surviving parts of the building or landscape, or that is physically, chemically, or ecologically incompatible.

Removing a feature of the building or landscape that is unrepairable and not replacing it; or replacing it with a new feature that does not convey the same visual appearance.

Setting 107

Rehabilitation

The following work is highlighted to indicate that it represents the particularly complex technical or design aspects of **Rehabilitation** *projects and should only be considered after the preservation concerns listed above have been addressed.*

Recommended	*Not Recommended*
Design for the Replacement of Missing Historic Features	
Designing and constructing a new feature of the building or landscape when the historic feature is completely missing, such as row house steps, a porch, a streetlight, or terrace. It may be a restoration based on documentary or physical evidence; or be a new design that is compatible with the historic character of the setting.	Creating a false historical appearance because the replaced feature is based on insufficient documentary or physical evidence. Introducing a new building or landscape feature that is out of scale or otherwise inappropriate to the setting's historic character, e.g., replacing picket fencing with chain link fencing.
Alterations/Additions for the New Use	
Designing required new parking so that it is as unobtrusive as possible, thus minimizing the effect on the historic character of the setting. "Shared" parking should also be planned so that several businesses can utilize one parking area as opposed to introducing random, multiple lots.	Placing parking facilities directly adjacent to historic buildings which result in damage to historic landscape features, such as the removal of plant material, relocation of paths and walkways, or blocking of alleys.
Designing and constructing new additions to historic buildings when required by the new use. New work should be compatible with the historic character of the setting in terms of size, scale design, material, color, and texture.	Introducing new construction into historic districts that is visually incompatible or that destroys historic relationships within the setting.
Removing nonsignificant buildings, additions or landscape features which detract from the historic character of the setting.	Removing a historic building, building feature, or landscape feature that is important in defining the historic character of the setting.

Rehabilitation

If a rear elevation of a historic building is distinctive and highly visible in the neighborhood, altering it may not meet the Standards. (a and b) This 3-story brick rowhouse featured a second story gallery and brick kitchen wing characteristic of other residences in the district which backed onto a connecting roadway. (c) In the rehabilitation, the wing and gallery were demolished and a large addition constructed that severely impacted the building's historic form and character.

Setting 109

Rehabilitation

Although the work in these sections is quite often an important aspect of rehabilitation projects, it is usually not part of the overall process of preserving character-defining features (maintenance, repair, replacement); rather, such work is assessed for its potential negative impact on the building's historic character. For this reason, particular care must be taken not to obscure, radically change, damage, or destroy character-defining features in the process of rehabilitation work.

Energy Efficiency

Recommended

Not Recommended

Masonry/Wood/Architectural Metals

Installing thermal insulation in attics and in unheated cellars and crawlspaces to increase the efficiency of the existing mechanical systems.

Installing insulating material on the inside of masonry walls to increase energy efficiency where there is no character-defining interior molding around the windows or other interior architectural detailing.

Applying thermal insulation with a high moisture content in wall cavities which may damage historic fabric.

Installing wall insulation without considering its effect on interior molding or other architectural detailing.

Windows

Utilizing the inherent energy conserving features of a building by maintaining windows and louvered blinds in good operable condition for natural ventilation.

Improving thermal efficiency with weatherstripping, storm windows, caulking, interior shades, and if historically appropriate, blinds and awnings.

Installing interior storm windows with air-tight gaskets, ventilating holes, and/or removable clips to ensure proper maintenance and to avoid condensation damage to historic windows.

Installing exterior storm windows which do not damage or obscure the windows and frames.

Removing historic shading devices rather than keeping them in an operable condition.

Replacing historic multi-paned sash with new thermal sash utilizing false muntins.

Installing interior storm windows that allow moisture to accumulate and damage the window.

Installing new exterior storm windows which are inappropriate in size or color.

Replacing windows or transoms with fixed thermal glazing or permitting windows and transoms to remain inoperable rather than utilizing them for their energy conserving potential.

Rehabilitation

Recommended	*Not Recommended*
Entrances and Porches	
Maintaining porches and double vestibule entrances so that they can retain heat or block the sun and provide natural ventilation.	Changing the historic appearance of the building by enclosing porches.
Interior Features	
Retaining historic interior shutters and transoms for their inherent energy conserving features.	Removing historic interior features which play an energy conserving role.
Mechanical Systems	
Improving energy efficiency of existing mechanical systems by installing insulation in attics and basements.	Replacing existing mechanical systems that could be repaired for continued use.
Building Site	
Retaining plant materials, trees, and landscape features which perform passive solar energy functions such as sun shading and wind breaks.	Removing plant materials, trees, and landscape features that perform passive solar energy functions.
Setting (District/Neighborhood)	
Maintaining those existing landscape features which moderate the effects of the climate on the setting such as deciduous trees, evergreen wind-blocks, and lakes or ponds.	Stripping the setting of landscape features and landforms so that effects of the wind, rain, and sun result in accelerated deterioration of the historic building.
New Additions to Historic Buildings	
Placing a new addition that may be necessary to increase energy efficiency on non-character-defining elevations.	Designing a new addition which obscures, damages, or destroys character-defining features.

Energy Efficiency

Rehabilitation

New Additions to Historic Buildings

Recommended

Placing functions and services required for the new use in non-character-defining interior spaces rather than constructing a new addition.

Constructing a new addition so that there is the least possible loss of historic materials and so that character-defining features are not obscured, damaged, or destroyed.

Designing a new addition in a manner that makes clear what is historic and what is new.

Not Recommended

Expanding the size of the historic building by constructing a new addition when the new use could be met by altering non-character-defining interior spaces.

Attaching a new addition so that the character-defining features of the historic building are obscured, damaged, or destroyed.

Duplicating the exact form, material, style, and detailing of the historic building in a new addition so that the new work appears to be part of the historic building.

Imitating a historic style or period of architecture in a new addition.

Rehabilitation, like Preservation, acknowledges a building's change over time; the retention and repair of existing historic materials and features is thus always recommended. However, unlike Preservation, the dual goal of Rehabilitation is to—respectfully—add to or alter a building in order to meet new use requirements. This downtown Chicago library was expanded in 1981 when additional space was required with light and humidity control for the rare book collection. The compatible 10-story wing was linked to the historic block on side and rear elevations. Its simple design is compatible with the historic form, features, and detailing; old and new are clearly differentiated. Photo: Dave Clifton.

Rehabilitation

Recommended

Considering the design for an attached exterior addition in terms of its relationship to the historic building as well as the historic district or neighborhood. Design for the new work may be contemporary or may reference design motifs from the historic building. In either case, it should always be clearly differentiated from the historic building and be compatible in terms of mass, materials, relationship of solids to voids, and color.

Placing a new addition on a non-character-defining elevation and limiting the size and scale in relationship to the historic building.

Designing a rooftop addition when required for the new use, that is set back from the wall plane and as inconspicuous as possible when viewed from the street.

Not Recommended

Designing and constructing new additions that result in the diminution or loss of the historic character of the resource, including its design, materials, workmanship, location, or setting.

Designing a new addition that obscures, damages, or destroys character-defining features of the historic building.

Constructing a rooftop addition so that the historic appearance of the building is radically changed.

Rehabilitation

Accessibility Considerations

Recommended

Identifying the historic building's character-defining spaces, features, and finishes so that accessibility code-required work will not result in their damage or loss.

Complying with barrier-free access requirements, in such a manner that character-defining spaces, features, and finishes are preserved.

Working with local disability groups, access specialists, and historic preservation specialists to determine the most appropriate solution to access problems.

Providing barrier-free access that promotes independence for the disabled person to the highest degree practicable, while preserving significant historic features.

Designing new or additional means of access that are compatible with the historic building and its setting.

Not Recommended

Undertaking code-required alterations before identifying those spaces, features, or finishes which are character-defining and must therefore be preserved.

Altering, damaging, or destroying character-defining features in attempting to comply with accessibility requirements.

Making changes to buildings without first seeking expert advice from access specialists and historic preservationists, to determine solutions.

Making access modifications that do not provide a reasonable balance between independent, safe access and preservation of historic features.

Designing new or additional means of access without considering the impact on the historic building and its setting.

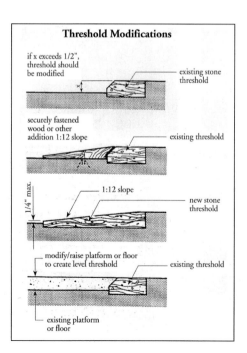

Making a building accessible to the public is a requirement under the Americans with Disabilities Act of 1990, whatever the treatment. Full, partial, or alternative approaches to accessibility depends upon the historical significance of a building and the ability to make changes. In these examples, thresholds that exceed allowable heights were modified several ways to increase accessibility, without jeopardizing the historic character. Drawing: Uniform Federal Accessibility Standard (UFAS) Retrofit Manual.

Rehabilitation

Health and Safety Considerations

Recommended

Identifying the historic building's character-defining spaces, features, and finishes so that code-required work will not result in their damage or loss.

Complying with health and safety codes, including seismic code requirements, in such a manner that character-defining spaces, features, and finishes are preserved.

Removing toxic building materials only after thorough testing has been conducted and only after less invasive abatement methods have been shown to be inadequate.

Providing workers with appropriate personal protective equipment for hazards found in the worksite.

Working with local code officials to investigate systems, methods, or devices of equivalent or superior effectiveness and safety to those prescribed by code so that unnecessary alterations can be avoided.

Upgrading historic stairways and elevators to meet health and safety codes in a manner that assures their preservation, i.e., so that they are not damaged or obscured.

Installing sensitively designed fire suppression systems, such as sprinkler systems that result in retention of historic features and finishes.

Applying fire-retardant coatings, such as intumescent paints, which expand during fire to add thermal protection to steel.

Adding a new stairway or elevator to meet health and safety codes in a manner that preserves adjacent character-defining features and spaces.

Placing a code-required stairway or elevator that cannot be accommodated within the historic building in a new exterior addition. Such an addition should be on an inconspicuous elevation.

Not Recommended

Undertaking code-required alterations to a building or site before identifying those spaces, features, or finishes which are character-defining and must therefore be preserved.

Altering, damaging, or destroying character-defining spaces, features, and finishes while making modifications to a building or site to comply with safety codes.

Destroying historic interior features and finishes without careful testing and without considering less invasive abatement methods.

Removing unhealthful building materials without regard to personal and environmental safety.

Making changes to historic buildings without first exploring equivalent health and safety systems, methods, or devices that may be less damaging to historic spaces, features, and finishes.

Damaging or obscuring historic stairways and elevators or altering adjacent spaces in the process of doing work to meet code requirements.

Covering character-defining wood features with fire-resistant sheathing which results in altering their visual appearance.

Using fire-retardant coatings if they damage or obscure character-defining features.

Radically changing, damaging, or destroying character-defining spaces, features, or finishes when adding a new code-required stairway or elevator.

Constructing a new addition to accommodate code-required stairs and elevators on character-defining elevations highly visible from the street; or where it obscures, damages, or destroys character-defining features.

Standards for Restoration & Guidelines for Restoring Historic Buildings

Restoration *is defined as the act or process of accurately depicting the form, features, and character of a property as it appeared at a particular period of time by means of the removal of features from other periods in its history and reconstruction of missing features from the restoration period. The limited and sensitive upgrading of mechanical, electrical, and plumbing systems and other code-required work to make properties functional is appropriate within a restoration project.*

Standards for Restoration

1. A property will be used as it was historically or be given a new use which reflects the property's restoration period.

2. Materials and features from the restoration period will be retained and preserved. The removal of materials or alteration of features, spaces, and spatial relationships that characterize the period will not be undertaken.

3. Each property will be recognized as a physical record of its time, place, and use. Work needed to stabilize, consolidate and conserve materials and features from the restoration period will be physically and visually compatible, identifiable upon close inspection, and properly documented for future research.

4. Materials, features, spaces, and finishes that characterize other historical periods will be documented prior to their alteration or removal.

5. Distinctive materials, features, finishes, and construction techniques or examples of craftsmanship that characterize the restoration period will be preserved.

6. Deteriorated features from the restoration period will be repaired rather than replaced. Where the severity of deterioration requires replacement of a distinctive feature, the new feature will match the old in design, color, texture, and, where possible, materials.

7. Replacement of missing features from the restoration period will be substantiated by documentary and physical evidence. A false sense of history will not be created by adding conjectural features, features from other properties, or by combining features that never existed together historically.

8. Chemical or physical treatments, if appropriate, will be undertaken using the gentlest means possible. Treatments that cause damage to historic materials will not be used.

9. Archeological resources affected by a project will be protected and preserved in place. If such resources must be disturbed, mitigation measures will be undertaken.

10. Designs that were never executed historically will not be constructed.

Guidelines for Restoring Historic Buildings

Introduction

Rather than maintaining and preserving a building as it has evolved over time, the expressed goal of the **Standards for Restoration and Guidelines for Restoring Historic Buildings** is to make the building appear as it did at a particular—and most significant—time in its history. First, those materials and features from the "restoration period" are identified, based on thorough historical research. Next, features from the restoration period are maintained, protected, repaired (i.e., stabilized, consolidated, and conserved), and replaced, if necessary. As opposed to other treatments, the scope of work in **Restoration** can include removal of features from other periods; missing features from the restoration period may be replaced, based on documentary and physical evidence, using traditional materials or compatible substitute materials. The final guidance emphasizes that only those designs that can be documented as having been built should be re-created in a restoration project.

Identify, Retain, and Preserve Materials and Features from the Restoration Period

The guidance for the treatment **Restoration** begins with recommendations to identify the form and detailing of those existing architectural materials and features that are significant to the restoration period as established by historical research and documentation. Thus, guidance on *identifying, retaining, and preserving* features from the restoration period is always given first. The historic building's appearance may be defined by the form and detailing of its exterior materials, such as masonry, wood, and metal; exterior features, such as roofs, porches, and windows; interior materials, such as plaster and paint; and interior features, such as moldings and stairways, room configuration and spatial relationships, as well as structural and mechanical systems; and the building's site and setting.

Protect and Maintain Materials and Features from the Restoration Period

After identifying those existing materials and features from the restoration period that must be retained in the process of **Restoration** work, then *protecting and maintaining* them is addressed. Protection generally involves the least degree of intervention and is preparatory to other work. For example, protection includes the maintenance of historic material through treatments such as rust removal, caulking, limited paint removal, and re-application of protective coatings; the cyclical cleaning of roof gutter systems; or installation of fencing, alarm systems and other temporary protective measures. Although a historic building will usually require more extensive work, an overall evaluation of its physical condition should always begin at this level.

Repair (Stabilize, Consolidate, and Conserve) Materials and Features from the Restoration Period

Next, when the physical condition of restoration period features requires additional work, *repairing* by *stabilizing, consolidating, and conserving* is recommended. **Restoration** guidance focuses upon the preservation of those materials and features that are significant to the period. Consequently, guidance for repairing a historic material, such as masonry, again begins with the least degree of intervention possible, such as strengthening fragile materials through consolidation, when appropriate, and repointing with mortar of an appropriate strength. Repairing masonry as well as wood and architectural metals includes patching, splicing, or otherwise reinforcing them using recognized preservation methods. Similarly, portions of a historic structural system could be reinforced using contemporary material such as steel rods. In **Restoration**, repair may also include the limited replacement in kind—or with compatible substitute material—of extensively deteriorated or missing parts of existing features when there are surviving prototypes to use as a model. Examples could include terra-cotta brackets, wood balusters, or cast iron fencing.

Replace Extensively Deteriorated Features from the Restoration Period

In **Restoration**, *replacing* an entire feature from the restoration period (i.e., a cornice, balustrade, column, or stairway) that is too deteriorated to repair may be appropriate. Together with documentary evidence, the form and detailing of the historic feature should be used as a model for the replacement. Using the same kind of material is preferred; however, compatible substitute material may be considered. All new work should be unobtrusively dated to guide future research and treatment.

In a project at Fort Hays, Kansas, the wood frame officers' quarters were restored to the late 1860s—their period of significance. This included replacing a missing kitchen ell, chimneys, porch columns, and cornice, and closing a later window opening in the main block. The building and others in the museum complex is used to interpret frontier history.

If documentary and physical evidence are not available to provide an accurate re-creation of missing features, the treatment Rehabilitation might be a better overall approach to project work.

Remove Existing Features from Other Historic Periods

Most buildings represent continuing occupancies and change over time, but in **Restoration**, the goal is to depict the building as it appeared at the most significant time in its history. Thus, work is included to remove or alter existing historic features that do not represent the restoration period. This could include features such as windows, entrances and doors, roof dormers, or landscape features. Prior to altering or removing materials, features, spaces, and finishes that characterize other historical periods, they should be documented to guide future research and treatment.

Re-Create Missing Features from the Restoration Period

Most **Restoration** projects involve re-creating features that were significant to the building at a particular time, but are now missing. Examples could include a stone balustrade, a porch, or cast iron storefront. Each missing feature should be substantiated by documentary and physical evidence. Without sufficient documentation for these "re-creations," an accurate depiction cannot be achieved. Combining features that never existed together historically can also create a false sense of history. Using traditional materials to depict lost features is always the preferred approach; however, using compatible substitute material is an acceptable alternative in **Restoration** because, as emphasized, the goal of this treatment is to replicate the "appearance" of the historic building at a particular time, not to retain and preserve all historic materials as they have evolved over time.

If documentary and physical evidence are not available to provide an accurate re-creation of missing features, the treatment Rehabilitation might be a better overall approach to project work.

Energy Efficiency/Accessibility Considerations/ Health and Safety Code Considerations

These sections of the **Restoration** guidance address work done to meet accessibility requirements and health and safety code requirements; or limited retrofitting measures to improve energy efficiency. Although this work is quite often an important aspect of restoration projects, it is usually not part of the overall process of protecting, stabilizing, conserving, or repairing features from the restoration period; rather, such work is assessed for its potential negative impact on the building's historic appearance. For this reason, particular care must be taken not to obscure, damage, or destroy historic materials or features from the restoration period in the process of undertaking work to meet code and energy requirements.

***Restoration as a Treatment.** When the property's design, architectural, or historical significance during a particular period of time outweighs the potential loss of extant materials, features, spaces, and finishes that characterize other historical periods; when there is substantial physical and documentary evidence for the work; and when contemporary alterations and additions are not planned, Restoration may be considered as a treatment. Prior to undertaking work, a particular period of time, i.e., the restoration period, should be selected and justified, and a documentation plan for Restoration developed.*

Restoration

Building Exterior

Masonry: Brick, stone, terra cotta, concrete, adobe, stucco and mortar

Recommended

Identifying, retaining, and preserving masonry features from the restoration period such as walls, brackets, railings, cornices, window architraves, door pediments, steps, and columns; and details such as tooling and bonding patterns, coatings, and color.

Protecting and maintaining masonry from the restoration period by providing proper drainage so that water does not stand on flat, horizontal surfaces or accumulate in curved decorative features.

Cleaning masonry only when necessary to halt deterioration or remove heavy soiling.

Carrying out masonry surface cleaning tests after it has been determined that such cleaning is appropriate. Tests should be observed over a sufficient period of time so that both the immediate and the long range effects are known to enable selection of the gentlest method possible.

Not Recommended

Altering masonry features from the restoration period.

Failing to properly document masonry features from the restoration period which may result in their loss.

Applying paint or other coatings such as stucco to masonry or removing paint or stucco from masonry if such treatments cannot be documented to the restoration period.

Changing the type or color of the paint or coating unless the work can be substantiated by historical documentation.

Failing to evaluate and treat the various causes of mortar joint deterioration such as leaking roofs or gutters, differential settlement of the building, capillary action, or extreme weather exposure.

Cleaning masonry surfaces when they are not heavily soiled, thus needlessly introducing chemicals or moisture into historic materials.

Cleaning masonry surfaces without testing or without sufficient time for the testing results to be of value.

Restoration

Recommended

Cleaning masonry surfaces with the gentlest method possible, such as low pressure water and detergents, using natural bristle brushes.

Inspecting painted masonry surfaces to determine whether repainting is necessary.

Removing damaged or deteriorated paint only to the next sound layer using the gentlest method possible (e.g., hand-scraping) prior to repainting.

Applying compatible paint coating systems following proper surface preparation.

Repainting with colors that are documented to the restoration period of the building.

Evaluating the existing condition of the masonry to determine whether more than protection and maintenance are required, that is, if repairs to masonry features from the restoration period will be necessary.

Repairing, stabilizing and conserving fragile masonry from the restoration period by well-tested consolidants, when appropriate. Repairs should be physically and visually compatible and identifiable upon close inspection for future research.

Not Recommended

Sandblasting brick or stone surfaces using dry or wet grit or other abrasives. These methods of cleaning permanently erode the surface of the material and accelerate deterioration.

Using a cleaning method that involves water or liquid chemical solutions when there is any possibility of freezing temperatures.

Cleaning with chemical products that will damage masonry, such as using acid on limestone or marble, or leaving chemicals on masonry surfaces.

Applying high pressure water cleaning methods that will damage historic masonry and the mortar joints.

Removing paint that is firmly adhering to, and thus protecting, masonry surfaces.

Using methods of removing paint which are destructive to masonry, such as sandblasting, application of caustic solutions, or high pressure waterblasting.

Failing to follow manufacturers' product and application instructions when repainting masonry.

Using new paint colors that are not documented to the restoration period of the building.

Failing to undertake adequate measures to assure the protection of masonry features from the restoration period.

Removing masonry from the restoration period that could be stabilized, repaired and conserved; or using untested consolidants and untrained personnel, thus causing further damage to fragile historic materials.

Building Exterior *Masonry*

Restoration

Recommended	*Not Recommended*
Repairing masonry walls and other masonry features by repointing the mortar joints where there is evidence of deterioration such as disintegrating mortar, cracks in mortar joints, loose bricks, damp walls, or damaged plasterwork.	Removing nondeteriorated mortar from sound joints, then repointing the entire building to achieve a uniform appearance.
Removing deteriorated mortar by carefully hand-raking the joints to avoid damaging the masonry.	Using electric saws and hammers rather than hand tools to remove deteriorated mortar from joints prior to repointing.
Duplicating and, if necessary, reproducing period mortar in strength, composition, color, and texture.	Repointing with mortar of high portland cement content (unless it is the content of the historic mortar). This can often create a bond that is stronger than the historic material and can cause damage as a result of the differing coefficient of expansion and the differing porosity of the material and the mortar.
	Repointing with a synthetic caulking compound.
	Using a "scrub" coating technique to repoint instead of traditional repointing methods.
Duplicating and, if necessary, reproducing period mortar joints in width and in joint profile.	Changing the width or joint profile when repointing.
Repairing stucco by removing the damaged material and patching with new stucco that duplicates stucco of the restoration period in strength, composition, color, and texture.	Removing sound stucco; or repairing with new stucco that is stronger than the historic material or does not convey the same visual appearance.
Using mud plaster as a surface coating over unfired, unstabilized adobe because the mud plaster will bond to the adobe.	Applying cement stucco to unfired, unstabilized adobe. Because the cement stucco will not bond properly, moisture can become entrapped between materials, resulting in accelerated deterioration of the adobe.
Cutting damaged concrete back to remove the source of deterioration (often corrosion on metal reinforcement bars). The new patch must be applied carefully so it will bond satisfactorily with, and match, the historic concrete.	Patching concrete without removing the source of deterioration.

Restoration

Recommended

Repairing masonry features from the restoration period by patching, piecing-in, or otherwise reinforcing the masonry using recognized preservation methods. Repair may also include the limited replacement in kind—or with compatible substitute material—of those extensively deteriorated or missing parts of masonry features from the restoration period when there are surviving prototypes such as terra-cotta brackets or stone balusters. The new work should be unobtrusively dated to guide future research and treatment.

Applying new or non-historic surface treatments such as water-repellent coatings to masonry only after repointing and only if masonry repairs have failed to arrest water penetration problems.

Not Recommended

Replacing an entire masonry feature from the restoration period such as a cornice or balustrade when repair of the masonry and limited replacement of deteriorated or missing parts are appropriate.

Using a substitute material for the replacement part that does not convey the visual appearance of the surviving parts of the masonry feature or that is physically or chemically incompatible.

Applying waterproof, water repellent, or non-historic coatings such as stucco to masonry as a substitute for repointing and masonry repairs. Coatings are frequently unnecessary, expensive, and may change the appearance of historic masonry as well as accelerate its deterioration.

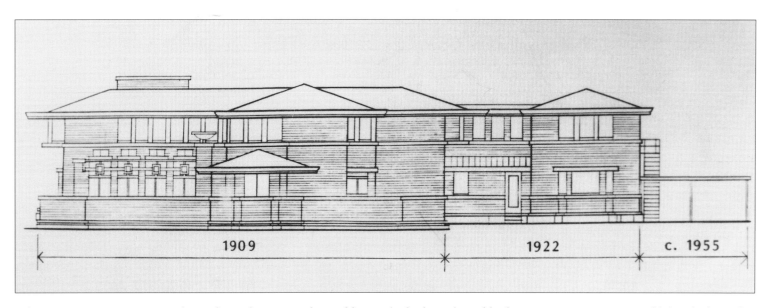

The Meyer May House in Grand Rapids, Michigan, was designed by Frank Lloyd Wright and built in 1909. In 1922, May added to the house for an expanding family. After the May occupancy, the house was altered for use as apartments, with a carport added in 1955. In the 1980s restoration, the Wright's original design was deemed more significant than May's later changes, and, as a result, the additions were removed and the house returned to its 1909 appearance. Drawing: Martha L. Werenfels, AIA.

Restoration

Recommended

Replacing in kind an entire masonry feature from the restoration period that is too deteriorated to repair—if the overall form and detailing are still evident—using the physical evidence as a model to reproduce the feature. Examples can include large sections of a wall, a cornice, balustrade, column, or stairway. If using the same kind of material is not technically or economically feasible, then a compatible substitute material may be considered. The new work should be unobtrusively dated to guide future research and treatment.

Not Recommended

Removing a masonry feature from the restoration period that is unrepairable and not replacing it.

The following **Restoration** *work is highlighted to indicate that it involves the removal or alteration of existing historic masonry features that would be retained in Preservation and Rehabilitation treatments; and the replacement of missing masonry features from the restoration period using all new materials.*

Recommended

Removing Existing Features from Other Historic Periods

Removing or altering masonry features from other historic periods such as a later doorway, porch, or steps.

Documenting materials and features dating from other periods prior to their alteration or removal. If possible, selected examples of these features or materials should be stored to facilitate future research.

Re-creating Missing Features from the Restoration Period

Re-creating a missing masonry feature that existed during the restoration period based on physical or documentary evidence; for example, duplicating a terra-cotta bracket or stone balustrade.

Not Recommended

Failing to remove a masonry feature from another period, thus confusing the depiction of the building's significance.

Failing to document masonry features from other historic periods that are removed from the building so that a valuable portion of the historic record is lost.

Constructing a masonry feature that was part of the original design for the building but was never actually built; or constructing a feature which was thought to have existed during the restoration period, but for which there is insufficient documentation.

Restoration

Building Exterior

Wood: Clapboard, weatherboard, shingles, and other wooden siding and decorative elements

Recommended

Identifying, retaining, and preserving wood features from the restoration period such as siding, cornices, brackets, window architraves, and doorway pediments; and their paints, finishes, and color.

Protecting and maintaining wood features from the restoration period by providing proper drainage so that water is not allowed to stand on flat, horizontal surfaces or accumulate in decorative features.

Applying chemical preservatives to wood features such as beam ends or outriggers that are exposed to decay hazards and are traditionally unpainted.

Retaining coatings such as paint that help protect the wood from moisture and ultraviolet light. Paint removal should be considered only where there is paint surface deterioration and as part of an overall maintenance program which involves repainting or applying other appropriate protective coatings.

Inspecting painted wood surfaces to determine whether repainting is necessary or if cleaning is all that is required.

Removing damaged or deteriorated paint to the next sound layer using the gentlest method possible (handscraping and handsanding), then repainting.

Not Recommended

Altering wood features from the restoration period.

Failing to properly document wood features from the restoration period which may result in their loss.

Applying paint or other coatings to wood or removing paint from wood if such treatments cannot be documented to the restoration period.

Changing the type or color of the paint or coating unless the work can be substantiated by historical documentation.

Failing to identify, evaluate, and treat the causes of wood deterioration, including faulty flashing, leaking gutters, cracks and holes in siding, deteriorated caulking in joints and seams, plant material growing too close to wood surfaces, or insect or fungus infestation.

Using chemical preservatives such as creosote which, unless they were used historically, can change the appearance of wood features.

Stripping paint or other coatings to reveal bare wood, thus exposing historically coated surfaces to the effects of accelerated weathering.

Removing paint that is firmly adhering to, and thus, protecting wood surfaces.

Using destructive paint removal methods such as propane or butane torches, sandblasting or waterblasting. These methods can irreversibly damage historic woodwork.

Restoration

Ongoing work at this house focuses on the maintenance and repair of exterior wood features from the restoration period. After scraping and sanding, the wood was painted in colors documented to the Restoration period. Photo: ©Mary Randlett, 1992.

Recommended

Using with care electric hot-air guns on decorative wood features and electric heat plates on flat wood surfaces when paint is so deteriorated that total removal is necessary prior to repainting.

Using chemical strippers primarily to supplement other methods such as handscraping, handsanding and the above-recommended thermal devices. Detachable wooden elements such as shutters, doors, and columns may—with the proper safeguards—be chemically dip-stripped.

Not Recommended

Using thermal devices improperly so that the historic woodwork is scorched.

Failing to neutralize the wood thoroughly after using chemicals so that new paint does not adhere.

Allowing detachable wood features to soak too long in a caustic solution so that the wood grain is raised and the surface roughened.

Restoration

Recommended

Applying compatible paint coating systems following proper surface preparation.

Repainting with colors that are documented to the restoration period of the building.

Evaluating the existing condition of the wood to determine whether more than protection and maintenance are required, that is, if repairs to wood features from the restoration period will be necessary.

Repairing, stabilizing, and conserving fragile wood from the restoration period using well-tested consolidants, when appropriate. Repairs should be physically and visually compatible and identifiable upon close inspection for future research.

Repairing wood features from the restoration period by patching, piecing-in, or otherwise reinforcing the wood using recognized preservation methods. Repair may also include the limited replacement in kind—or with compatible substitute material—of those extensively deteriorated or missing parts of features from the restoration period where there are surviving prototypes such as brackets, molding, or sections of siding. The new work should be unobtrusively dated to guide future research and treatment.

Replacing in kind an entire wood feature from the restoration period that is too deteriorated to repair—if the overall form and detailing are still evident—using the physical evidence as a model to reproduce the feature. Examples of wood features include a cornice, entablature or balustrade. If using the same kind of material is not technically or economically feasible, then a compatible substitute material may be considered. The new work should be unobtrusively dated to guide future research and treatment.

Not Recommended

Failing to follow manufacturers' product and application instructions when repainting exterior woodwork.

Using new colors that are not documented to the restoration period of the building.

Failing to undertake adequate measures to assure the protection of wood features from the restoration period.

Removing wood from the restoration period that could be stabilized and conserved; or using untested consolidants and untrained personnel, thus causing further damage to fragile historic materials.

Replacing an entire wood feature from the restoration period such as a cornice or wall when repair of the wood and limited replacement of deteriorated or missing parts are appropriate.

Using substitute material for the replacement part that does not convey the visual appearance of the surviving parts of the wood feature or that is physically or chemically incompatible.

Removing a wood feature from the restoration period that is unrepairable and not replacing it.

Restoration

The following **Restoration** *work is highlighted to indicate that it involves the removal or alteration of existing historic wood features that would be retained in Preservation and Rehabilitation treatments; and the replacement of missing wood features from the restoration period using all new materials.*

Recommended	*Not Recommended*
Removing Existing Features from Other Historic Periods	
Removing or altering wood features from other historic periods such as a later doorway, porch, or steps.	Failing to remove a wood feature from another period, thus confusing the depiction of the building's significance.
Documenting materials and features dating from other periods prior to their alteration or removal. If possible, selected examples of these features or materials should be stored to facilitate future research.	Failing to document wood features from other historic periods that are removed from the building so that a valuable portion of the historic record is lost.
Re-creating Missing Features from the Restoration Period	
Re-creating a missing wood feature that existed during the restoration period based on physical or documentary evidence; for example, duplicating a roof dormer or porch.	Constructing a wood feature that was part of the original design for the building, but was never actually built; or constructing a feature which was thought to have existed during the restoration period, but for which there is insufficient documentation.

Restoration

Building Exterior

Architectural Metals: Cast iron, steel pressed tin, copper, aluminum, and zinc

Recommended

Identifying, retaining, and preserving architectural metal features from the restoration period such as columns, capitals, window hoods, or stairways; and their finishes and colors. Identification is also critical to differentiate between metals prior to work. Each metal has unique properties and thus requires different treatments.

Protecting and maintaining restoration period architectural metals from corrosion by providing proper drainage so that water does not stand on flat, horizontal surfaces or accumulate in curved, decorative features.

Cleaning architectural metals, when appropriate, to remove corrosion prior to repainting or applying other appropriate protective coatings.

Identifying the particular type of metal prior to any cleaning procedure and then testing to assure that the gentlest cleaning method possible is selected or determining that cleaning is inappropriate for the particular metal.

Cleaning soft metals such as lead, tin, copper, terneplate, and zinc with appropriate chemical methods because their finishes can be easily abraded by blasting methods.

Not Recommended

Altering architectural metal features from the restoration period.

Failing to properly document architectural metal features from the restoration period which may result in their loss.

Changing the type of finish, historic color, or accent scheme unless the work can be substantiated by historical documentation.

Failing to identify, evaluate, and treat the causes of corrosion, such as moisture from leaking roofs or gutters.

Exposing metals which were intended to be protected from the environment.

Applying paint or other coatings to metals such as copper, bronze, or stainless steel that were meant to be exposed.

Using cleaning methods which alter or damage the historic color, texture, and finish of the metal; or cleaning when it is inappropriate for the metal.

Removing the patina of historic metal. The patina may be a protective coating on some metals, such as bronze or copper, as well as a significant historic finish.

Cleaning soft metals such as lead, tin, copper, terneplate, and zinc with grit blasting which will abrade the surface of the metal.

Restoration

Recommended

Using the gentlest cleaning methods for cast iron, wrought iron, and steel—hard metals—in order to remove paint buildup and corrosion. If handscraping and wire brushing have proven ineffective, low pressure grit blasting may be used as long as it does not abrade or damage the surface.

Applying appropriate paint or other coating systems after cleaning in order to decrease the corrosion rate of metals or alloys.

Repainting with colors that are documented to the restoration period of the building.

Applying an appropriate protective coating such as lacquer to an architectural metal feature such as a bronze door which is subject to heavy pedestrian use.

Evaluating the existing condition of the architectural metals to determine whether more than protection and maintenance are required, that is, if repairs to metal features from the restoration period will be necessary.

Repairing, stabilizing, and conserving fragile architectural metal from the restoration period using well-tested consolidants, when appropriate. Repairs should be physically and visually compatible and identifiable upon close inspection for future research.

Repairing architectural metal features from the restoration period by patching, splicing, or otherwise reinforcing the metal using recognized preservation methods. Repairs may also include the limited replacement in kind—or with a compatible substitute material—of those extensively deteriorated or missing parts of features from the restoration period when there are surviving prototypes such as porch balusters, column capitals or bases; or porch cresting. The new work should be unobtrusively dated to guide future research and treatment.

Not Recommended

Failing to employ gentler methods prior to abrasively cleaning cast iron, wrought iron or steel; or using high pressure grit blasting.

Failing to re-apply protective coating systems to metals or alloys that require them after cleaning so that accelerated corrosion occurs.

Using new colors that are not documented to the restoration period of the building.

Failing to assess pedestrian use or new access patterns so that architectural metal features are subject to damage by use or inappropriate maintenance such as salting adjacent sidewalks.

Failing to undertake adequate measures to assure the protection of architectural metal features from the restoration period.

Removing architectural metal from the restoration period that could be stabilized and conserved; or using untested consolidants and untrained personnel, thus causing further damage to fragile historic materials.

Replacing an entire architectural metal feature from the restoration period such as a column or a balustrade when repair of the metal and limited replacement of deteriorated or missing parts are appropriate.

Using a substitute material for the replacement part that does not convey the visual appearance of the surviving parts of the architectural metal feature or that is physically or chemically incompatible.

Restoration

The Standards for Restoration call for the repair of existing features from the restoration period as well as the re-creation of missing features from the period. In some instances, when missing features are replaced, substitute materials may be considered if they convey the appearance of the historic materials. In this example at Philadelphia's Independence Hall, the clock was re-built in 1972-73 using cast stone and wood with fiberglass and polyester bronze ornamentation. Photo: Lee H. Nelson, FAIA.

Recommended

Replacing in kind an entire architectural metal feature from the restoration period that is too deteriorated to repair—if the overall form and detailing are still evident—using the physical evidence as a model to reproduce the feature. Examples could include cast iron porch steps or roof cresting. If using the same kind of material is not technically or economically feasible, then a compatible substitute material may be considered. The new work should be unobtrusively dated to guide future research and treatment.

Not Recommended

Removing an architectural metal feature from the restoration period that is unrepairable and not replacing it.

Building Exterior *Metals*

Restoration

The following **Restoration** *work is highlighted to indicate that it involves the removal or alteration of existing historic architectural metal features that would be retained in Preservation and Rehabilitation treatments; and the replacement of missing architectural metal features from the restoration period using all new materials.*

Recommended	*Not Recommended*
Removing Existing Features from Other Historic Periods	
Removing or altering architectural metal features from other historic periods such as a later cast iron porch railing or aluminum windows.	Failing to remove an architectural metal feature from another period, thus confusing the depiction of the building's significance.
Documenting materials and features dating from other periods prior to their alteration or removal. If possible, selected examples of these features or materials should be stored to facilitate future research.	Failing to document architectural metal features from other historic periods that are removed from the building so that a valuable portion of the historic record is lost.
Re-creating Missing Features from the Restoration Period	
Re-creating a missing architectural metal feature that existed during the restoration period based on physical or documentary evidence; for example, duplicating a cast iron storefront or porch.	Constructing an architectural metal feature that was part of the original design for the building but was never actually built; or constructing a feature which was thought to have existed during the restoration period, but for which there is insufficient documentation.

Restoration

Building Exterior

Roofs

Recommended

Identifying, retaining, and preserving roofs and roof features from the restoration period. This includes the roof's shape, such as hipped, gambrel, and mansard; decorative features such as cupolas, cresting, chimneys, and weathervanes; and roofing material such as slate, wood, clay tile, and metal, as well as size, color, and patterning.

Protecting and maintaining a restoration period roof by cleaning the gutters and downspouts and replacing deteriorated flashing. Roof sheathing should also be checked for proper venting to prevent moisture condensation and water penetration; and to insure that materials are free from insect infestation.

Providing adequate anchorage for roofing material to guard against wind damage and moisture penetration.

Protecting a leaking roof with plywood and building paper until it can be properly repaired.

Evaluating the existing condition of materials to determine whether more than protection and maintenance are required, that is, if repairs to roofs and roof features will be necessary.

Repairing a roof from the restoration period by reinforcing the materials which comprise roof features. Repairs will also generally include the limited replacement in kind—or with compatible substitute material—of those extensively deteriorated or missing parts of features when there are surviving prototypes such as cupola louvers, dentils, dormer roofing; or slates, tiles, or wood shingles. The new work should be unobtrusively dated to guide future research and treatment.

Not Recommended

Altering roofs and roof features from the restoration period.

Failing to properly document roof features from the restoration period which may result in their loss.

Changing the type or color of roofing materials unless the work can be substantiated by historical documentation.

Failing to clean and maintain gutters and downspouts properly so that water and debris collect and cause damage to roof fasteners, sheathing, and the underlying structure.

Allowing roof fasteners, such as nails and clips, to corrode so that roofing material is subject to accelerated deterioration.

Permitting a leaking roof to remain unprotected so that accelerated deterioration of historic building materials—masonry, wood, plaster, paint and structural members—occurs.

Failing to undertake adequate measures to assure the protection of roofs and roof features from the restoration period.

Replacing an entire roof feature from the restoration period such as a cupola or dormer when the repair of materials and limited replacement of deteriorated or missing parts are appropriate.

Failing to reuse intact slate or tile when only the roofing substrate needs replacement.

Using a substitute material for the replacement part that does not convey the visual appearance of the surviving parts of the roof or that is physically or chemically incompatible.

Restoration

Recommended

Replacing in kind an entire roof feature from the restoration period that is too deteriorated to repair—if the overall form and detailing are still evident—using the physical evidence as a model to reproduce the feature. Examples can include a large section of roofing, or a dormer or chimney. If using the same kind of material is not technically or economically feasible, then a compatible substitute material may be considered. The new work should be unobtrusively dated to guide future research and treatment.

Not Recommended

Removing a roof feature from the restoration period that is unrepairable, and not replacing it; or failing to document the new work.

The following **Restoration** *work involves the removal or alteration of existing historic roofs and roof features that would be retained in Preservation and Rehabilitation treatments; and the replacement of missing roof features from the restoration period using all new materials in order to create an accurate historic appearance.*

Recommended

Removing Existing Features from Other Historic Periods

Removing or altering roofs or roof features from other historic periods such as a later dormer or asphalt roofing.

Documenting materials and features dating from other periods prior to their alteration or removal. If possible, selected examples of these features or materials should be stored to facilitate future research.

Re-creating Missing Features from the Restoration Period

Re-creating missing roofing material or a roof feature that existed during the restoration period based on physical or documentary evidence; for example, duplicating a dormer or cupola.

Not Recommended

Failing to remove a roof feature from another period, thus confusing the depiction and of the building's significance.

Failing to document roofing materials and roof features from other historic periods that are removed from the building so that a valuable portion of the historic record is lost.

Constructing a roof feature that was part of the original design for the building, but was never actually built; or constructing a feature which was thought to have existed during the restoration period, but for which there is insufficient documentation.

Restoration

Building Exterior

Windows

Recommended

Identifying, retaining, and preserving windows—and their functional and decorative features—from the restoration period. Such features can include frames, sash, muntins, glazing, sills, heads, hoodmolds, panelled or decorated jambs and moldings, and interior and exterior shutters and blinds.

Conducting an indepth survey of the condition of existing windows from the restoration period early in the planning process so that repair and upgrading methods and possible replacement options can be fully explored.

Protecting and maintaining the wood and architectural metals from the restoration period which comprise the window frame, sash, muntins, and surrounds through appropriate surface treatments such as cleaning, rust removal, limited paint removal, and re-application of protective coating systems.

Making windows weathertight by re-caulking, and replacing or installing weatherstripping. These actions also improve thermal efficiency.

Evaluating the existing condition of materials to determine whether more than protection and maintenance are required, i.e. if repairs to windows and window features will be required.

Not Recommended

Altering windows or window features from the restoration period.

Failing to properly document window features from the restoration period which may result in their loss.

Applying paint or other coatings to window features or removing them if such treatments cannot be documented to the restoration period.

Changing the type or color of protective surface coatings on window features unless the work can be substantiated by historical documentation.

Stripping windows of sound material such as wood, cast iron, and bronze.

Replacing windows from the restoration period solely because of peeling paint, broken glass, stuck sash, and high air infiltration. These conditions, in themselves, are no indication that windows are beyond repair.

Failing to provide adequate protection of materials on a cyclical basis so that deterioration of the window results.

Retrofitting or replacing windows from the restoration period rather than maintaining the sash, frame, and glazing.

Failing to undertake adequate measures to assure the protection of window materials from the restoration period.

Restoration

Recommended

Repairing window frames and sash from the restoration period by patching, splicing, consolidating or otherwise reinforcing. Such repair may also include replacement in kind—or with compatible substitute material—of those extensively deteriorated or missing parts when there are surviving prototypes such as architraves, hoodmolds, sash, sills, and interior or exterior shutters and blinds. The new work should be unobtrusively dated to guide future research and treatment.

Replacing in kind a window feature from the restoration period that is too deteriorated to repair using the same sash and pane configuration and other design details. If using the same kind of material is not technically or economically feasible when replacing windows deteriorated beyond repair, then a compatible substitute material may be considered. The new work should be unobtrusively dated to guide future research and treatment.

Not Recommended

Replacing an entire window from the restoration period when repair of materials and limited replacement of deteriorated or missing parts are appropriate.

Failing to reuse serviceable window hardware such as brass sash lifts and sash locks.

Using a substitute material for the replacement part that does not convey the visual appearance of the surviving parts of the window or that is physically or chemically incompatible.

Removing a window feature from the restoration period that is unrepairable and not replacing it; or failing to document the new work.

Restoration

The following **Restoration** *work is highlighted to indicate that it involves the removal or alteration of existing historic windows and window features that would be retained in Preservation and Rehabilitation treatments; and the replacement of missing window features from the restoration period using all new materials.*

Recommended	*Not Recommended*
Removing Existing Features from Other Historic Periods	
Removing or altering windows or window features from other historic periods, such as later single-pane glazing or inappropriate shutters.	Failing to remove a window feature from another period, thus confusing the depiction of the building's significance.
Documenting materials and features dating from other periods prior to their alteration or removal. If possible, selected examples of these features or materials should be stored to facilitate future research.	Failing to document window features from other historic periods that are removed from the building so that a valuable portion of the historic record is lost.
Re-creating Missing Features from the Restoration Period	
Re-creating a missing window or window feature that existed during the restoration period based on physical or documentary evidence; for example, duplicating a hoodmold or shutter.	Constructing a window feature that was part of the original design for the building, but was never actually built; or constructing a feature which was thought to have existed during the restoration period, but for which there is insufficient documentation.

Restoration

Building Exterior

Entrances and Porches

Recommended	*Not Recommended*
Identifying, retaining, and preserving entrances and porches from the restoration period—and their functional and decorative features—such as doors, fanlights, sidelights, pilasters, entablatures, columns, balustrades, and stairs.	Altering entrances and porch features from the restoration period.
	Failing to properly document entrance and porch features from the restoration period which may result in their loss
	Applying paint or other coatings to entrance and porch features or removing them if such treatments cannot be documented to the restoration period.
	Changing the type or color of protective surface coatings on entrance and porch features unless the work can be substantiated by historical documentation.
	Stripping entrances and porches of sound material such as wood, iron, cast iron, terra cotta, tile and brick.
Protecting and maintaining the masonry, wood, and architectural metals that comprise restoration period entrances and porches through appropriate surface treatments such as cleaning, rust removal, limited paint removal, and re-application of protective coating systems.	Failing to provide adequate protection to materials on a cyclical basis so that deterioration of entrances and porches results.
Evaluating the existing condition of materials to determine whether more than protection and maintenance are required, that is, if repairs to entrance and porch features will be necessary.	Failing to undertake adequate measures to assure the protection of historic entrances and porches from the restoration period.

Restoration

Recommended

Repairing entrances and porches from the restoration period by reinforcing the historic materials. Repairs will also generally include the limited replacement in kind—or with compatible substitute material—of those extensively deteriorated or missing parts of repeated features where there are surviving prototypes such as balustrades, cornices, entablatures, columns, sidelights, and stairs. The new work should be unobtrusively dated to guide future research and treatment.

Replacing in kind an entire entrance or porch from the restoration period that is too deteriorated to repair—if the form and detailing are still evident—using the physical evidence as a model to reproduce the feature. If using the same kind of material is not technically or economically feasible, then a compatible substitute material may be considered. The new work should be unobtrusively dated to guide future research and treatment.

Not Recommended

Replacing an entire entrance or porch feature from the restoration period when the repair of materials and limited replacement of parts are appropriate.

Using a substitute material for the replacement part that does not convey the visual appearance of the surviving parts of the entrance and porch or that is physically or chemically incompatible.

Removing an entrance or porch feature from the restoration period that is unrepairable and not replacing it; or failing to document the new work.

Portions of the small porch on an Italianate mansion were carefully numbered prior to Restoration. Some original elements were restored in place, while others had to be removed for repair, then reinstalled. Any element too deteriorated to save was replaced with a new one replicated to match the original design. Photo: Morgan W. Phillips.

Building Exterior *Entrances and Porches*

Restoration

The following **Restoration** *work is highlighted to indicate that it involves the removal or alteration of existing historic entrance and porch features that would be retained in Preservation and Rehabilitation treatments; and the replacement of missing entrance and porch features from the restoration period using all new materials.*

Recommended	*Not Recommended*
Removing Existing Features from Other Historic Periods	
Removing or altering entrances and porches and their features from other historic periods such as a later porch railing or balustrade.	Failing to remove an entrance or porch feature from another period, thus confusing the depiction of the building's significance.
Documenting materials and features dating from other periods prior to their alteration or removal. If possible, selected examples of these features or materials should be stored to facilitate future research.	Failing to document entrance or porch features from other historic periods that are removed from the building so that a valuable portion of the historic record is lost.
Re-creating Missing Features from the Restoration Period	
Re-creating a missing entrance or porch or its features that existed during the restoration period based on physical or documentary evidence; for example, duplicating a fanlight or porch column.	Constructing an entrance or porch feature that was part of the original design for the building but was never actually built; or constructing a feature which was thought to have existed during the restoration period, but for which there is insufficient documentation.

Restoration

Building Exterior

Storefronts

Recommended

Identifying, retaining, and preserving storefronts from the restoration period—and their functional and decorative features—such as display windows, signs, doors, transoms, kick plates, corner posts, and entablatures.

Protecting and maintaining masonry, wood, and architectural metals which comprise restoration period storefronts through appropriate treatments such as cleaning, rust removal, limited paint removal, and reapplication of protective coating systems.

Protecting storefronts against arson and vandalism before restoration work begins by boarding up windows and installing alarm systems that are keyed into local protection agencies.

Evaluating the existing condition of storefront materials to determine whether more than protection and maintenance are required, that is, if repairs to features will be necessary.

Not Recommended

Altering storefronts—and their features—from the restoration period.

Failing to properly document storefront features from the restoration period which may result in their loss.

Applying paint or other coatings to storefront features or removing them if such treatments cannot be documented to the restoration period.

Changing the type or color of protective surface coatings on storefront features unless the work can be substantiated by historical documentation.

Failing to provide adequate protection of materials on a cyclical basis so that deterioration of storefront features results.

Permitting entry into the building through unsecured or broken windows and doors so that interior features and finishes are damaged by exposure to weather or vandalism.

Stripping storefronts of historic material from the restoration period such as wood, cast iron, terra cotta, carrara glass, and brick.

Failing to undertake adequate measures to assure the protection of storefront materials from the restoration period.

Restoration

Recommended

Repairing storefronts from the restoration period by reinforcing the historic materials. Repairs will also generally include the limited replacement in kind—or with compatible substitute materials—of those extensively deteriorated or missing parts of storefronts where there are surviving prototypes such as transoms, kick plates, pilasters, or signs. The new work should be unobtrusively dated to guide future research and treatment.

Replacing in kind a storefront from the restoration period that is too deteriorated to repair—if the overall form and detailing are still evident—using the physical evidence as a model. If using the same material is not technically or economically feasible, then compatible substitute materials may be considered. The new work should be unobtrusively dated to guide future research and treatment.

Not Recommended

Replacing an entire storefront feature from the restoration period when repair of materials and limited replacement of its parts are appropriate.

Using substitute material for the replacement part that does not convey the same visual appearance as the surviving parts of the storefront or that is physically or chemically incompatible.

Removing a storefront feature from the restoration period that is unrepairable, and not replacing it; or failing to document the new work.

The following **Restoration** *work is highlighted to indicate that it involves the removal or alteration of existing historic storefront features that would be retained in Preservation and Rehabilitation treatments; and the replacement of missing storefront features from the restoration period using all new materials.*

Recommended

Removing Existing Features from Other Historic Periods

Removing or altering storefronts and their features from other historic periods such as inappropriate cladding or signage.

Documenting materials and features dating from other periods prior to their alteration or removal. If possible, selected examples of these features or materials should be stored to facilitate future research.

Re-creating Missing Features from the Restoration Period

Re-creating a missing storefront or storefront feature that existed during the restoration period based on physical or documentary evidence; for example, duplicating a display window or transom.

Not Recommended

Failing to remove a storefront feature from another period, thus confusing the depiction of the building's significance.

Failing to document storefront features from other historic periods that are removed from the building so that a valuable portion of the historic record is lost.

Constructing a storefront feature that was part of the original design for the building but was never actually built; or constructing a feature which was thought to have existed during the restoration period, but for which there is insufficient documentation.

Building Interior
Structural Systems

Recommended

Identifying, retaining, and preserving structural systems from the restoration period—and individual features of systems—such as post and beam systems, trusses, summer beams, vigas, cast iron columns, above-grade stone foundation walls, or loadbearing brick or stone walls.

Protecting and maintaining the structural system by cleaning the roof gutters and downspouts; replacing roof flashing; keeping masonry, wood, and architectural metals in a sound condition; and ensuring that structural members are free from insect infestation.

Examining and evaluating the physical condition of the structural system and its individual features using non-destructive techniques such as X-ray photography.

Repairing the structural system by augmenting or upgrading individual parts or features in a manner that is consistent with the restoration period. For example, weakened structural members such as floor framing can be paired with a new member, braced, or otherwise supplemented and reinforced. The new work should be unobtrusively dated to guide future research and treatment.

Not Recommended

Altering visible features of structural systems from the restoration period.

Failing to properly document structural systems from the restoration period which may result in their loss.

Overloading the existing structural system; or installing equipment or mechanical systems which could damage the structure.

Replacing a loadbearing masonry wall that could be augmented and retained.

Leaving known structural problems untreated such as deflection of beams, cracking and bowing of walls, or racking of structural members.

Failing to provide proper building maintenance so that deterioration of the structural system results. Causes of deterioration include subsurface ground movement, vegetation growing too close to foundation walls, improper grading, fungal rot, and poor interior ventilation that results in condensation.

Utilizing destructive probing techniques that will damage or destroy structural material.

Upgrading the building structurally in a manner that diminishes the historic character of the exterior, such as installing strapping channels or removing a decorative cornice; or that damages interior features or spaces.

Replacing a structural member or other feature of the structural system when it could be augmented and retained.

Restoration

Recommended

Replacing in kind—or with substitute material—those portions or features of the structural system that are either extensively deteriorated or are missing when there are surviving prototypes such as cast iron columns, roof rafters or trusses, or sections of loadbearing walls. Substitute material should convey the same form, design, and overall visual appearance as the historic feature; and, at a minimum, be equal to its loadbearing capabilities. The new work should be unobtrusively dated to guide future research and treatment.

Not Recommended

Installing a visible replacement feature that does not convey the same visual appearance, e.g., replacing an exposed wood summer beam with a steel beam; or failing to document the new work.

Using substitute material that does not equal the loadbearing capabilities of the historic material and design or is otherwise physically or chemically incompatible.

The following **Restoration** *work is highlighted to indicate that it involves the removal or alteration of existing historic structural systems and features that would be retained in Preservation and Rehabilitation treatments; and the replacement of missing structural system features from the restoration period using all new materials.*

Recommended

Removing Existing Features from Other Historic Periods

Removing or altering visually intrusive structural features from other historic periods such as a non-matching column or exposed ceiling beams.

Documenting materials and features dating from other periods prior to their alteration or removal. If possible, selected examples of these features or materials should be stored to facilitate future research.

Re-creating Missing Features from the Restoration Period

Re-creating a missing structural feature that existed during the restoration period based on physical or documentary evidence; for example, duplicating a viga or cast iron column.

Not Recommended

Failing to remove or alter a visually intrusive structural feature from another period, thus confusing the depiction of the building's significance.

Failing to document structural features from other historic periods that are removed from the building so that a valuable portion of the historic record is lost.

Constructing a structural feature that was part of the original design for the building but was never actually built; or constructing a feature which was thought to have existed during the restoration period, but for which there is insufficient documentation.

Restoration

Building Interior

Spaces, Features, and Finishes

Recommended

Interior Spaces

Identifying, retaining, and preserving a floor plan or interior spaces from the restoration period. This includes the size, configuration, proportion, and relationship of rooms and corridors; the relationship of features to spaces; and the spaces themselves, such as lobbies, reception halls, entrance halls, double parlors, theaters, auditoriums, and important industrial or commercial spaces.

Interior Features and Finishes

Identifying, retaining, and preserving interior features and finishes from the restoration period. These include columns, cornices, baseboards, fireplaces and mantels, panelling, light fixtures, hardware, and flooring; and wallpaper, plaster, paint, and finishes such as stencilling, marbling, and graining; and other decorative materials that accent interior features and provide color, texture, and patterning to walls, floors, and ceilings.

Protecting and maintaining masonry, wood, and architectural metals that comprise restoration period interior features through appropriate surface treatments such as cleaning, rust removal, limited paint removal, and reapplication of protective coating systems.

Not Recommended

Altering a floor plan or interior spaces—including individual rooms—from the restoration period.

Altering features or finishes from the restoration period.

Failing to properly document spaces, features, and finishes from the restoration period which may result in their loss.

Applying paint, plaster, or other finishes to surfaces unless the work can be substantiated historical documentation.

Stripping paint to bare wood rather than repairing or reapplying grained or marbled finishes from the restoration period to features such as doors and panelling.

Changing the type of finish or its color, such as painting a previously varnished wood feature, unless the work can be substantiated by historical documentation.

Failing to provide adequate protection to materials on a cyclical basis so that deterioration of interior features results.

Restoration

Recommended

Protecting interior spaces, features and finishes against arson and vandalism before project work begins, erecting protective fencing, boarding-up windows, and installing fire alarm systems that are keyed to local protection agencies.

Protecting interior features such as a staircase, mantel, or decorative finishes and wall coverings against damage during project work by covering them with heavy canvas or plastic sheets.

Installing protective coverings in areas of heavy pedestrian traffic to protect historic features such as wall coverings, parquet flooring and panelling.

Removing damaged or deteriorated paints and finishes to the next sound layer using the gentlest method possible, then repainting or refinishing using compatible paint or other coating systems based on historical documentation.

Repainting with colors that are documented to the building's restoration period.

Limiting abrasive cleaning methods to certain industrial warehouse buildings where the interior masonry or plaster features do not have distinguishing design, detailing, tooling, or finishes; and where wood features are not finished, molded, beaded, or worked by hand. Abrasive cleaning should only be considered after other, gentler methods have been proven ineffective.

Evaluating the existing condition of materials to determine whether more than protection and maintenance are required, that is, if repairs to interior features and finishes will be necessary.

Not Recommended

Permitting entry into historic buildings through unsecured or broken windows and doors so that the interior features and finishes are damaged by exposure to weather or vandalism.

Stripping interiors of restoration period features such as woodwork, doors, windows, light fixtures, copper piping, radiators; or of decorative materials.

Failing to provide proper protection of interior features and finishes during work so that they are gouged, scratched, dented, or otherwise damaged.

Failing to take new use patterns into consideration so that interior features and finishes are damaged.

Using destructive methods such as propane or butane torches or sandblasting to remove paint or other coatings. These methods can irreversibly damage the historic materials that comprise interior features.

Using new paint colors that are inappropriate to the building's restoration period.

Changing the texture and patina of features from the restoration period through sandblasting or use of abrasive methods to remove paint, discoloration or plaster. This includes both exposed wood (including structural members) and masonry.

Failing to undertake adequate measures to assure the protection of interior features and finishes.

Recommended

Repairing interior features and finishes from the restoration period by reinforcing the historic materials. Repair will also generally include the limited replacement in kind—or with compatible substitute material—of those extensively deteriorated or missing parts of repeated features when there are surviving prototypes such as stairs, balustrades, wood panelling, columns; or decorative wall coverings or ornamental tin or plaster ceilings. The new work should be unobtrusively dated to guide future research and treatment.

Replacing in kind an entire interior feature or finish from the restoration period that is too deteriorated to repair—if the overall form and detailing are still evident—using the physical evidence as a model for reproduction. Examples could include wainscoting, a tin ceiling, or interior stairs. If using the same kind of material is not technically or economically feasible, then a compatible substitute material may be considered. The new work should be unobtrusively dated to guide future research and treatment.

Not Recommended

Replacing an interior feature from the restoration period such as a staircase, panelled wall, parquet floor, or cornice; or finish such as a decorative wall covering or ceiling when repair of materials and limited replacement of such parts are appropriate.

Using a substitute material for the replacement part that does not convey the visual appearance of the surviving parts or portions of the interior feature or finish or that is physically or chemically incompatible.

Removing a feature or finish from the restoration period that is unrepairable and not replacing it; or failing to document the new work.

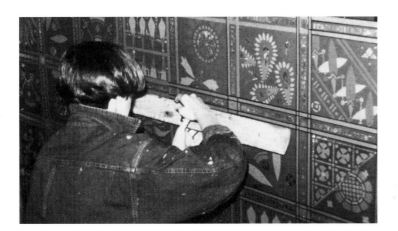
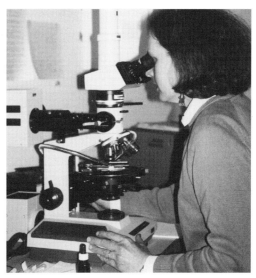

A complete paint investigation often needs to be conducted during Restoration. Paint samples are carefully collected onsite. In the laboratory, an ultra violet light is used to identify pigment and binding media. Paint samples are then photographed. Physical evidence documented through laboratory research provides a sound basis for an accurate restoration of painted finishes, such as the complex stencilling pictured here. Photo left: Courtesy, Alexis Elza; Photo right: Courtesy, Andrea Gilmore.

Restoration

The following **Restoration** *work is highlighted to indicate that it involves the removal or alteration of existing historic interior spaces, features, and finishes that would be retained in Preservation and Rehabilitation treatments; and the replacement of missing interior spaces, features, and finishes from the restoration period using all new materials.*

Recommended

Removing Existing Features from Other Historic Periods

Removing or altering interior spaces, features and finishes from other historic periods such as a later suspended ceiling or wood panelling.

Documenting materials and features dating from other periods prior to their alteration or removal. If possible, selected examples of these features or materials should be stored to facilitate future research.

Re-creating Missing Features from the Restoration Period

Re-creating an interior space, or a missing feature or finish from the restoration period based on physical or documentary evidence; for example, duplicating a marbleized mantel or a staircase.

Not Recommended

Failing to remove or alter an interior space, feature, or finish from another period, thus confusing the depiction of the building's significance.

Failing to document interior spaces, features, and finishes from other historic periods that are removed from the building so that a valuable portion of the historic record is lost.

Constructing an interior space, feature, or finish that was part of the original design for the building but was never actually built; or constructing a feature which was thought to have existed during the restoration period, but for which there is insufficient documentation.

The missing plaster cornice was restored as part of an overall project to return a residence to its original appearance. The traditional method of producing a cornice is unchanged today. Photo: Old-House Journal.

Restoration

Building Interior

Mechanical Systems: Heating, Air Conditioning, Electrical, and Plumbing

Recommended

Identifying, retaining, and preserving visible features of mechanical systems from the restoration period such as radiators, vents, fans, grilles, plumbing fixtures, switchplates, and lights.

Protecting and maintaining mechanical, plumbing, and electrical systems and their features from the restoration period through cyclical cleaning and other appropriate measures.

Preventing accelerated deterioration of mechanical systems by providing adequate ventilation of attics, crawlspaces, and cellars so that moisture problems are avoided.

Improving the energy efficiency of existing mechanical systems to help reduce the need for elaborate new equipment.

Repairing mechanical systems from the restoration period by augmenting or upgrading system parts, such as installing new pipes and ducts; rewiring; or adding new compressors or boilers.

Replacing in kind—or with compatible substitute material—those visible features of restoration period mechanical systems that are either extensively deteriorated or are prototypes such as ceiling fans, switchplates, radiators, grilles, or plumbing fixtures.

Installing a new mechanical system, if required, in a way that results in the least alteration possible to the building.

Not Recommended

Altering visible decorative features of mechanical systems from the restoration period.

Failing to properly document mechanical systems and their visible decorative features from the restoration period which may result in their loss.

Failing to provide adequate protection of materials on a cyclical basis so that deterioration of mechanical systems and their visible features results.

Enclosing mechanical systems in areas that are not adequately ventilated so that deterioration of the systems results.

Installing unnecessary air conditioning or climate control systems which can add excessive moisture to the building. This additional moisture can either condense inside, damaging interior surfaces, or pass through interior walls to the exterior, potentially damaging adjacent materials as it migrates.

Replacing a mechanical system from the restoration period or its functional parts when it could be upgraded and retained.

Installing a visible replacement feature that does not convey the same visual appearance.

Installing a new mechanical system so that structural or interior features from the restoration period are altered.

Restoration

Recommended

Providing adequate structural support for new mechanical equipment.

Installing the vertical runs of ducts, pipes, and cables in closets, service rooms, and wall cavities.

Installing air conditioning units in such a manner that features are not damaged or obscured and excessive moisture is not generated that will accelerate deterioration of historic materials.

Not Recommended

Failing to consider the weight and design of new mechanical equipment so that, as a result, historic structural members or finished surfaces are weakened or cracked.

Installing vertical runs of ducts, pipes, and cables in places where they will obscure features from the restoration period.

Concealing mechanical equipment in walls or ceilings in a manner that requires the removal of building material from the restoration period.

Cutting through features such as masonry walls in order to install air conditioning units.

The following **Restoration** *work is highlighted to indicate that it involves the removal or alteration of existing historic mechanical systems and features that would be retained in Preservation and Rehabilitation treatments; and the replacement of missing mechanical systems and features from the restoration period using all new materials.*

Recommended

Removing Existing Features from Other Historic Periods

Removing or altering mechanical systems and features from other historic periods such as a later elevator or plumbing fixture.

Documenting materials and features dating from other periods prior to their alteration or removal. If possible, selected examples of these features or materials should be stored to facilitate future research.

Re-creating Missing Features from the Restoration Period

Re-creating a missing feature of the mechanical system that existed during the restoration period based on physical or documentary evidence; for example, duplicating a heating vent or gaslight fixture.

Not Recommended

Failing to remove a mechanical system or feature from another period, thus confusing the depiction of the building's significance.

Failing to document mechanical systems and features from other historic periods that are removed from the building so that a valuable portion of the historic record is lost.

Constructing a mechanical system or feature that was part of the original design for the building but was never actually built; or constructing a feature which was thought to have existed during the restoration period, but for which there is insufficient documentation.

Restoration

Building Site

Recommended

Identifying, retaining, and preserving restoration period buildings and their features as well as features of the site. Site features may include circulation systems such as walks, paths, roads, or parking; vegetation such as trees, shrubs, fields, or herbaceous plant material; landforms such as terracing, berms or grading; furnishings such as lights, fences, or benches; decorative elements such as sculpture, statuary or monuments; water features including fountains, streams, pools, or lakes; and subsurface archeological features which are important in defining the restoration period.

Not Recommended

Altering buildings and their features or site features from the restoration period.

Failing to properly document building and site features from the restoration period which may result in their loss.

This ca. 1900 photograph (left) would be invaluable to guide restoration of the deteriorated house (right) to its documented earlier appearance, complete with decorative trim, shutters, polychromed exterior, and fencing. Photos: Courtesy, North Carolina Department of Archives and History.

Restoration

Recommended

Re-establishing the relationship between buildings and the landscape that existed during the restoration period.

Protecting and maintaining buildings and the site by providing proper drainage to assure that water does not erode foundation walls; drain toward the building; or damage or erode the landscape.

Minimizing disturbance of terrain around buildings or elsewhere on the site, thus reducing the possibility of destroying or damaging important landscape features or archeological resources.

Surveying and documenting areas where the terrain will be altered during restoration work to determine the potential impact to landscape features or archeological resources.

Protecting, e.g., preserving in place, important archeological resources.

Planning and carrying out any necessary investigation using professional archeologists and modern archeological methods when preservation in place is not feasible.

Preserving important landscape features from the restoration period, including ongoing maintenance of historic plant material.

Protecting building and landscape features against arson and vandalism before restoration work begins, i.e., erecting protective fencing and installing alarm systems that are keyed into local protection agencies.

Not Recommended

Retaining non-restoration period buildings or landscape features.

Failing to maintain adequate site drainage so that buildings and site features are damaged or destroyed; or alternatively, changing the site grading so that water no longer drains properly.

Introducing heavy machinery into areas where it may disturb or damage important landscape features or archeological resources.

Failing to survey the building site prior to beginning restoration work which results in damage to, or destruction of, landscape features or archeological resources.

Leaving known archeological material unprotected so that it is damaged during restoration work.

Permitting unqualified personnel to perform data recovery on archeological resources so that improper methodology results in the loss of important archeological material.

Allowing restoration period landscape features to be lost or damaged due to a lack of maintenance.

Permitting the property to remain unprotected so that the building and landscape features or archeological resources are damaged or destroyed.

Removing restoration period features from the building or site such as wood siding, iron fencing, masonry balustrades, or plant material.

Restoration

Recommended

Providing continued protection of building materials and plant features from the restoration period through appropriate cleaning, rust removal, limited paint removal, and re-application of protective coating systems; and pruning and vegetation management.

Evaluating the existing condition of materials and features to determine whether more than protection and maintenance are required, that is, if repairs to building and site features will be necessary.

Repairing restoration period features of the building and site by reinforcing historic materials. The new work should be unobtrusively dated to guide future research and treatment.

Replacing in kind an entire restoration period feature of the building or site that is too deteriorated to repair if the overall form and detailing are still evident. Physical evidence from the deteriorated feature should be used as a model to guide the new work. This could include an entrance or porch, walkway, or fountain. If using the same kind of material is not technically or economically feasible, then a compatible substitute material may be considered. The new work should be unobtrusively dated to guide future research and treatment.

Replacing deteriorated or damaged landscape features of the restoration period in kind or with compatible substitute material. The replacement feature should be based on physical evidence and convey the same appearance.

Not Recommended

Failing to provide adequate protection of materials on a cyclical basis so that deterioration of building and site features results.

Failing to undertake adequate measures to assure the protection of building and site features.

Replacing an entire restoration period feature of the building or site such as a fence, walkway, or driveway when repair of materials and limited compatible replacement of deteriorated or missing parts are appropriate.

Using a substitute material for the replacement part that does not convey the visual appearance of the surviving parts of the building or site feature or that is physically or chemically incompatible.

Removing a restoration period feature of the building or site that is unrepairable and not replacing it; or failing to document the new work.

Adding conjectural landscape features to the site such as period reproduction lamps, fences, fountains, or vegetation that are historically inappropriate, thus creating an inaccurate depiction of the restoration period.

Building Site 155

Restoration

*The following **Restoration** work is highlighted to indicate that it involves the removal or alteration of existing historic building site features that would be retained in Preservation and Rehabilitation treatments; and the replacement of missing building site features from the restoration period using all new materials.*

Recommended	*Not Recommended*
Removing Existing Features from Other Historic Periods	
Removing or altering features of the building or site from other historic periods such as a later outbuilding, paved road, or overgrown tree.	Failing to remove a feature of the building or site from another period, thus creating an inaccurate historic appearance.
Documenting features of the building or site from other periods prior to their alteration or removal.	Failing to document features of the building or site from other historic periods that are removed during restoration so that a valuable portion of the historic record is lost.
Re-creating Missing Features from the Restoration Period	
Re-creating a missing feature of the building or site that existed during the restoration period based on physical or documentary evidence; for example, duplicating a terrace, gazebo, or fencing.	Constructing a feature of the building or site that was part of the original design, but was never actually built; or constructing a feature which was thought to have existed during the restoration period, but for which there is insufficient documentation.

Restoration

Setting (District/Neighborhood)

Recommended

Identifying retaining, and preserving restoration period building and landscape features of the setting. Such features can include roads and streets, furnishings such as lights or benches, vegetation, gardens and yards, adjacent open space such as fields, parks, commons or woodlands, and important views or visual relationships.

Re-establishing the relationship between buildings and landscape features of the setting that existed during the restoration period.

Protecting and maintaining building materials and plant features from the restoration period through appropriate cleaning, rust removal, limited paint removal, and reapplication of protective coating systems; and pruning and vegetation management.

Protecting buildings and landscape features against arson and vandalism before restoration work begins by erecting protective fencing and installing alarm systems that are keyed into local protection agencies.

Evaluating the existing condition of the building and landscape features to determine whether more than protection and maintenance are required, that is, if repairs to features will be necessary.

Repairing restoration period features of the building and landscape by reinforcing the historic materials. Repair will generally include the replacement in kind—or with compatible substitute material—of those extensively deteriorated or missing parts of features where there are surviving prototypes such as porch balustrades or paving materials. The new work should be unobtrusively dated to guide future research and treatment.

Not Recommended

Altering features of the setting that can be documented to the restoration period.

Failing to properly document restoration period building and landscape features, which may result in their loss.

Retaining non-restoration period buildings or landscape features.

Failing to provide adequate protection of materials on a cyclical basis which results in the deterioration of building and landscape features.

Permitting the building and setting to remain unprotected so that interior or exterior features are damaged.

Stripping or removing features from buildings or the setting such as wood siding, iron fencing, terra cotta balusters, or plant material.

Failing to undertake adequate measures to assure the protection of building and landscape features.

Replacing an entire restoration period feature of the building or landscape setting when repair of materials and limited replacement of deteriorated or missing parts are appropriate.

Using a substitute material for the replacement part that does not convey the visual appearance of the surviving parts of the building or landscape, or that is physically, chemically, or ecologically incompatible.

Setting 157

Restoration

Recommended

Replacing in kind an entire restoration period feature of the building or landscape that is too deteriorated to repair—when the overall form and detailing are still evident—using the physical evidence as a model to guide the new work. If using the same kind of material is not technically or economically feasible, then a compatible substitute material may be considered. The new work should be unobtrusively dated to guide future research and treatment.

Not Recommended

Removing a restoration period feature of the building or landscape that is unrepairable and not replacing it; or failing to document the new work.

*The following **Restoration** work is highlighted to indicate that it involves the removal or alteration of existing features of the historic setting that would be retained in Preservation and Rehabilitation treatments; and the replacement of missing features from the restoration period using all new materials.*

Recommended

Removing Existing Features from Other Historic Periods

Removing or altering features of the building or landscape from other historic periods, such as a later road, sidewalk, or fence.

Documenting features of the building or landscape dating from other periods prior to their alteration or removal.

Re-creating Missing Features from the Restoration Period

Re-creating a missing feature of the building or landscape in the setting that existed during the restoration period based on physical or documentary evidence; for example, duplicating a path or park bench.

Not Recommended

Failing to remove a feature of the building or landscape from another period, thus creating an inaccurate historic appearance.

Failing to document features of the building or landscape from other historic periods that are removed from the setting so that a valuable portion of the historic record is lost.

Constructing a feature of the building or landscape that was part of the original design for the setting but was never actually built; or constructing a feature which was thought to have existed during the restoration period, but for which there is insufficient documentation.

Restoration

The Bronson-Mulholland House in Palatka, Florida, ca. 1845, is shown (a) before and (b) after the treatment, Restoration. Over the years the east (far right) side of the veranda had been filled in as a sixth bay. During the restoration, this later infill was removed and the east veranda, together with its flooring, stairs, and foundation, restored. Photo: City of Palatka, Community Development Department.

Setting

Restoration

Although the work in the following sections is quite often an important aspect of restoration projects, it is usually not part of the overall process of preserving features from the restoration period (protection, stabilization, conservation, repair, and replacement); rather, such work is assessed for its potential negative impact on the building's historic appearance. For this reason, particular care must be taken not to obscure, alter, or damage features from the restoration period in the process of undertaking work to meet code and energy requirements.

Energy Efficiency

Recommended	*Not Recommended*
Masonry/Wood/Architectural Metals	
Installing thermal insulation in attics and in unheated cellars and crawlspaces to increase the efficiency of the existing mechanical systems.	Applying thermal insulation with a high moisture content in wall cavities which may damage historic fabric.
Installing insulating material on the inside of masonry walls to increase energy efficiency where there is no interior molding around the windows or other interior architectural detailing from the restoration period.	Installing wall insulation without considering its effect on interior or other architectural detailing.
Windows	
Utilizing the inherent energy conserving features of a building by maintaining windows and louvered blinds from the restoration period in good operable condition for natural ventilation.	Using shading devices that are inappropriate to the restoration period.
Improving thermal efficiency with weatherstripping, storm windows, caulking, interior shades, and if historically appropriate, blinds and awnings.	Replacing multi-paned sash from the restoration period with new thermal sash utilizing false muntins.
Installing interior storm windows with air-tight gaskets, ventilating holes, and/or removable clips to ensure proper maintenance and to avoid condensation damage to historic windows.	Installing interior storm windows that allow moisture to accumulate and damage the window.
Installing exterior storm windows which do not damage or obscure the windows and frames.	Installing new exterior storm windows which are inappropriate in size or color.
	Replacing windows or transoms from the restoration period with fixed thermal glazing or permitting windows and transoms to remain inoperable rather than utilizing them for their energy conserving potential.

Restoration

Recommended	*Not Recommended*

Entrances and Porches

Maintaining porches and double vestibule entrances from the restoration period so that they can retain heat or block the sun and provide natural ventilation.

Changing porches significant to the restoration period by enclosing them.

Interior Features

Retaining interior shutters and transoms from the restoration period for their inherent energy conserving features.

Removing interior features from the restoration period that play a secondary energy conserving role.

Mechanical Systems

Improving energy efficiency of existing mechanical systems by installing insulation in attics and basements.

Replacing existing mechanical systems that could be repaired for continued use.

Building Site

Retaining plant materials, trees, and landscape features which perform passive solar energy functions, such as sun shading and wind breaks, if appropriate to the restoration period.

Removing plant materials, trees, and landscape features from the restoration period that perform passive solar energy functions.

Setting (District/Neighborhood)

Maintaining those existing landscape features which moderate the effects of the climate on the setting such as deciduous trees, evergreen wind-blocks, and lakes or ponds, if appropriate to the restoration period.

Stripping the setting of landscape features and landforms from the restoration period so that effects of the wind, rain, and sun result in accelerated deterioration of the historic building.

Restoration

Accessibility Considerations

Recommended

Identifying spaces, features, and finishes from the restoration period so that accessibility code-required work will not result in their damage or loss.

Complying with barrier-free access requirements in such a manner that spaces, features, and finishes from the restoration period are preserved.

Working with local disability groups, access specialists, and historic preservation specialists to determine the most appropriate solution to access problems.

Providing barrier-free access that promotes independence for to the highest degree practicable, while preserving significant historic features.

Finding solutions to meet accessibility requirements that minimize the impact on the historic building and its site, such as compatible ramps, paths, and lifts.

Not Recommended

Undertaking code-required alterations before identifying those spaces, features, or finishes from the restoration period which must be preserved.

Altering, damaging, or destroying features from the restoration period in attempting to comply with accessibility requirements.

Making changes to buildings without first seeking expert advice from access specialists and historic preservationists to determine solutions.

Making access modifications that do not provide a reasonable balance between independent, safe access and preservation of historic features.

Making modifications for accessibility without considering the impact on the historic building and its site.

Restoration

Health and Safety Considerations

Recommended

Identifying spaces, features, and finishes from the restoration period so that code-required work will not result in their damage or loss.

Complying with health and safety codes, including seismic code requirements, in such a manner that spaces, features, and finishes from the restoration period are preserved.

Removing toxic building materials only after thorough testing has been conducted and only after less invasive abatement methods have been shown to be inadequate.

Providing workers with appropriate personal protective equipment for hazards found at the worksite.

Working with local code officials to investigate systems, methods, or devices of equivalent or superior effectiveness and safety to those prescribed by code so that unnecessary alterations can be avoided.

Upgrading historic stairways and elevators from the restoration period to meet health and safety codes in a manner that assures their preservation, i.e., so that they are not damaged or obscured.

Installing sensitively designed fire suppression systems, such as sprinkler systems, that result in retention of features and finishes from the restoration period.

Applying fire-retardant coatings, such as intumescent paints, which expand during fire to add thermal protection to steel.

Adding a new stairway or elevator to meet health and safety codes in a manner that preserves adjacent features and spaces from the restoration period.

Not Recommended

Undertaking code-required alterations to a building or site before identifying those spaces, features, or finishes from the restoration period which must be preserved.

Altering, damaging, or destroying spaces, features, and finishes while making modifications to a building or site to comply with safety codes.

Destroying interior features and finishes from the restoration period without careful testing and without considering less invasive abatement methods.

Removing unhealthful building materials without regard to personal and environmental safety.

Making changes to historic buildings without first exploring equivalent health and safety systems, methods, or devices that may be less damaging to spaces, features, and finishes from the restoration period.

Damaging or obscuring stairways and elevators or altering adjacent spaces from the restoration period in the process of doing work to meet code requirements.

Covering wood features from the restoration period with fire-resistant sheathing which results in altering their visual appearance.

Using fire-retardant coatings if they damage or obscure features from the restoration period.

Altering the appearance of spaces, features, or finishes from the restoration period when adding a new code-required stairway or elevator.

Standards for Reconstruction & Guidelines for Reconstructing Historic Buildings

Reconstruction *is defined as the act or process of depicting, by means of new construction, the form, features, and detailing of a non-surviving site, landscape, building, structure, or object for the purpose of replicating its appearance at a specific period of time and in its historic location.*

Standards for Reconstruction

1. Reconstruction will be used to depict vanished or non-surviving portions of a property when documentary and physical evidence is available to permit accurate reconstruction with minimal conjecture, and such reconstruction is essential to the public understanding of the property.

2. Reconstruction of a landscape, building, structure, or object in its historic location will be preceded by a thorough archeological investigation to identify and evaluate those features and artifacts which are essential to an accurate reconstruction. If such resources must be disturbed, mitigation measures will be undertaken.

3. Reconstruction will include measures to preserve any remaining historic materials, features, and spatial relationships.

4. Reconstruction will be based on the accurate duplication of historic features and elements substantiated by documentary or physical evidence rather than on conjectural designs or the availability of different features from other historic properties. A reconstructed property will re-create the appearance of the non-surviving historic property in materials, design, color, and texture.

5. A reconstruction will be clearly identified as a contemporary re-creation.

6. Designs that were never executed historically will not be constructed.

Guidelines for Reconstructing Historic Buildings

Introduction

Whereas the treatment Restoration provides guidance on restoring—or re-creating—building features, the **Standards for Reconstruction and Guidelines for Reconstructing Historic Buildings** address those aspects of treatment necessary to re-create an entire non-surviving building with new material. Much like restoration, the goal is to make the building appear as it did at a particular—and most significant—time in its history. The difference is, in **Reconstruction**, there is far less extant historic material prior to treatment and, in some cases, nothing visible. Because of the potential for historical error in the absence of sound physical evidence, this treatment can be justified only rarely and, thus, is the least frequently undertaken. Documentation requirements prior to and following work are very stringent. Measures should be taken to preserve extant historic surface and subsurface material. Finally, the reconstructed building must be clearly identified as a contemporary re-creation.

In the 1930s reconstruction of the 18th century Governor's Palace at Colonial Williamsburg, Virginia, the archeological remains of the brick foundation were carefully preserved in situ, and serve as a base for the reconstructed walls. Photo: The Colonial Williamsburg Foundation.

Research and Document Historical Significance

Guidance for the treatment **Reconstruction** begins with *researching and documenting* the building's historical significance to ascertain that its re-creation is essential to the public understanding of the property. Often, another extant historic building on the site or in a setting can adequately explain the property, together with other interpretive aids. Justifying a reconstruction requires detailed physical and documentary evidence to minimize or eliminate conjecture and ensure that the reconstruction is as accurate as possible. Only one period of significance is generally identified; a building, as it evolved, is rarely re-created. During this important fact-finding stage, if research does not provide adequate documentation for an accurate reconstruction, other interpretive methods should be considered, such as an explanatory marker.

Investigate Archeological Resources

Investigating archeological resources is the next area of guidance in the treatment **Reconstruction**. The goal of physical research is to identify features of the building and site which are essential to an accurate re-creation and must be reconstructed, while leaving those archeological resources that are not essential, undisturbed. Information that is not relevant to the project should be preserved in place for future research. The archeological findings, together with archival documentation, are then used to replicate the plan of the building, together with the relationship and size of rooms, corridors, and other spaces, and spatial relationships.

Identify, Protect and Preserve Extant Historic Features

Closely aligned with archeological research, recommendations are given for *identifying, protecting, and preserving* extant features of the historic building. It is never appropriate to base a **Reconstruction** upon conjectural designs or the availability of different features from other buildings. Thus, any remaining historic materials and features, such as remnants of a foundation or chimney and site features such as a walkway or path, should be retained, when practicable, and incorporated into the reconstruction. The historic as well as new material should be carefully documented to guide future research and treatment.

Reconstruct Non-Surviving Building and Site

After the research and documentation phases, guidance is given for **Reconstruction** work itself. Exterior and interior features are addressed in general, always emphasizing the need for an accurate *depiction*, i.e., careful duplication of the appearance of historic interior paints, and finishes such as stencilling, marbling, and graining. In the absence of extant historic materials, the objective in reconstruction is to re-create the appearance of the historic building for interpretive purposes. Thus, while the use of traditional materials and finishes is always preferred, in some instances, substitute materials may be used if they are able to convey the same visual appearance.

Where non-visible features of the building are concerned—such as interior structural systems or mechanical systems—it is expected that contemporary materials and technology will be employed.

Re-creating the building site should be an integral aspect of project work. The initial archeological inventory of subsurface and aboveground remains is used as documentation to reconstruct landscape features such as walks and roads, fences, benches, and fountains.

Energy Efficiency/Accessibility/Health and Safety Code Considerations

Code requirements must also be met in Reconstruction projects. For code purposes, a reconstructed building may be considered as essentially new construction. Guidance for these sections is thus abbreviated, and focuses on achieving design solutions that do not destroy extant historic features and materials or obscure reconstructed features.

***Reconstruction as a Treatment.** When a contemporary depiction is required to understand and interpret a property's historic value (including the re-creation of missing components in a historic district or site); when no other property with the same associative value has survived; and when sufficient historical documentation exists to ensure an accurate reproduction, Reconstruction may be considered as a treatment. Prior to undertaking work, a documentation plan for Reconstruction should be developed.*

Reconstruction should generally be based on an extensive archeological investigation, as was done here to re-create a non-surviving commissary building at Fort Snelling.

Reconstruction

Recommended

Researching and documenting the property's historical significance, focusing on the availability of documentary and physical evidence needed to justify reconstruction of the non-surviving building.

Investigating archeological resources to identify and evaluate those features and artifacts which are essential to the design and plan of the building.

Not Recommended

Undertaking a reconstruction based on insufficient research, so that, as a result, an historically inaccurate building is created.

Reconstructing a building unnecessarily when an existing building adequately reflects or explains the history of the property, the historical event, or has the same associative value.

Executing a design for the building that was never constructed historically.

Failing to identify and evaluate archeological information prior to reconstruction, or destroying extant historical information not relevant to the reconstruction but which should be preserved in place.

Jean Baptiste Wengler's watercolor rendering of Fort Snelling, Minnesota, in 1857, is aesthetically pleasing, but the overall view does not constitute adequate documentary evidence for a Reconstruction. Oral histories are also unreliable sources of documentation for treatment.

Reconstruction

Recommended

Minimizing disturbance of terrain to reduce the possibility of destroying archeological resources.

Identifying, retaining, and preserving extant historic features of the building and site, such as remnants of a foundation, chimney, or walkway.

Not Recommended

Introducing heavy machinery or equipment into areas where it may disturb archeological resources.

Beginning reconstruction work without first conducting a detailed site investigation to physically substantiate the documentary evidence.

Basing a reconstruction on conjectural designs or the availability of different features from other historic buildings.

(a) and (b). Two photos illustrate the use of contemporary construction materials and techniques within the treatment, Reconstruction. Because Reconstruction is employed to portray a significant earlier time, usually for interpretive purposes, substitute materials may be appropriate if they are able to convey the historic appearance.

Reconstruction

Recommended

Building Exterior

Reconstructing a non-surviving building to depict the documented historic appearance. Although traditional building materials such as masonry, wood, and architectural metals are preferable, substitute materials may be used as long as they re-create the historical appearance.

Re-creating the documented design of exterior features such as the roof shape and coverings; architectural detailing; windows; entrances and porches; steps and doors; and their historic spatial relationships and proportions.

Reproducing the appearance of historic paint colors and finishes based on physical and documentary evidence.

Using signs to identify the building as a contemporary re-creation.

Building Interior

Re-creating the appearance of *visible* features of the historical structural system, such as post and beam systems, trusses, summer beams, vigas, cast iron columns, above-grade stone foundations, or loadbearing brick or stone walls. Substitute materials may be used for unexposed structural features if they were not important to the historic significance of the building.

Re-creating a historic floor plan or interior spaces, including the size, configuration, proportion, and relationship of rooms and corridors; the relationship of features to spaces; and the spaces themselves.

Not Recommended

Reconstructing features that cannot be documented historically or for which inadequate documentation exists.

Using substitute materials that do not convey the appearance of the historic building.

Omitting a documented exterior feature; or re-building a feature, but altering its historic design.

Using inappropriate designs or materials that do not convey the historic appearance, such as aluminum storm and screen window combinations.

Using paint colors that cannot be documented through research and investigation to be appropriate to the building or using other undocumented finishes.

Failing to explain that the building is a reconstruction, thus confusing the public understanding.

Changing the documented appearance of visible features of the structural system.

Altering the documented historic floor plan or relocating an important interior feature such as a staircase so that the historic relationship between the feature and space is inaccurately depicted.

Reconstruction

Recommended

Duplicating the documented historic appearance of the building's interior features and finishes, including columns, cornices, baseboards, fireplaces and mantels, panelling, light fixtures, hardware, and flooring; and wallpaper, plaster, paint and finishes such as stencilling, marbling and graining; and other decorative materials that accented interior features and provided color, texture, and patterning to walls, floors and ceilings.

Installing modern mechanical systems in the least obtrusive way possible, while meeting user need.

Installing the vertical runs of ducts, pipes, and cables in closets, service rooms, and wall cavities.

Installing exterior electrical and telephone cables underground, or in the least obtrusive way possible.

Not Recommended

Altering the documented appearance of interior features and finishes so that, as a result, an inaccurate depiction of the historic building is created. For example, moving a feature from one area of a room to another; or changing the type or color of the finish.

Altering the historic plan or the re-created appearance unnecessarily when installing modern mechanical systems.

Installing vertical runs in ducts, pipes, and cables in places where they will intrude upon the historic depiction of the building.

Attaching exterior electrical and telephone cables to the principal elevations of the reconstructed building, unless their existence and visibility can be documented.

Building Interior

Reconstruction

The spacious grounds at Middleton Place, near Charleston, South Carolina, constitute the first landscaped garden in America. The molded terraces, originally constructed in the 18th century, were largely reconstructed in the early 20th century based on extant remains and other documentary evidence. Photo: Middleton Place.

Building Site

Recommended

Basing decisions for reconstructing building site features on the availability of documentary and physical evidence.

Inventorying the building site to determine the existence of aboveground remains and subsurface archeological materials, then using this evidence as corroborating documentation for the reconstruction of related site features. These may include walks, paths, roads, and parking; trees, shrubs, fields or herbaceous plant material; terracing, berms, or grading; lights, fences, or benches; sculpture, statuary, or monuments; fountains, streams, pools, or lakes.

Re-establishing the historic relationship between the building or buildings and historic site features, whenever possible.

Not Recommended

Reconstructing building site features without first conducting a detailed investigation to physically substantiate the documentary evidence.

Giving the building's site a false appearance by basing the reconstruction or conjectural designs or the availability of features from other nearby sites.

Changing the historic spatial relationship between the building and historic site features, or reconstructing some site features, but not others, thus creating a false appearance.

Reconstruction

Recommended

Setting (District or Neighborhood)

Basing decisions for reconstructing features of the building's setting on the availability of documentary and physical evidence.

Inventorying the setting to determine the existence of aboveground remains and subsurface archeological materials, using this evidence as corroborating documentation for the reconstruction of missing features of the setting. Such features could include roads and streets; furnishings such as lights or benches; vegetation, gardens and yards; adjacent open space such as fields, parks, commons or woodlands; and important views or visual relationships.

Re-establishing the historic spatial relationship between buildings and landscape features of the setting.

Not Recommended

Reconstructing features of the setting without first conducting a detailed investigation to physically substantiate the documentary evidence.

Giving the building's setting a false appearance by basing the reconstruction on conjectural designs or the availability of features from other nearby districts or neighborhoods.

Confusing the historic spatial relationship between buildings and landscape features within the setting by reconstructing some missing elements, but not others.

(a) and (b). Two views of the Officers' Quarters at Fort Snelling (ca. 1885 to 1900) not only provide information on the materials and form of the historic block, they document the wooden walkways and other landscape features such as stairs, railings, and tree placement. Historical and pictorial evidence would need to be combined with specific physical evidence in order to make the case for Reconstruction as a treatment.

Reconstruction

The 1778 Kershaw House, which served as British Headquarters during the Revolutionary War, was burned by Union troops in 1865. In the early 1970s, the house was reconstructed as part of Camden Battlefield, Camden, South Carolina. Built expressly for interpretive purposes, it serves as an illustrative reminder of a past event of national significance. The Standards for Reconstruction call for any re-created building to be clearly identified as a contemporary depiction. This is most often done by means of an exterior sign or plaque, or through an explanatory brochure or exhibit. A guide may inform visitors as well. Photo: Richard Frear.

Reconstruction

*Whereas preservation, rehabilitation, and restoration treatments usually necessitate retrofitting to meet code and energy requirements, in this treatment it is assumed that the reconstructed building will be essentially **new** construction. Thus, only minimal guidance is provided in the following section, although the work must still be assessed for its potential negative impact on the reconstructed*

Recommended

Energy Efficiency

Installing thermal insulation, where appropriate, as part of the reconstruction.

Utilizing the inherent energy conserving features of windows and blinds, porches and double vestibule entrances in a reconstruction project.

Utilizing plant materials, trees, and landscape features, especially those which perform passive solar energy functions such as sun shading and wind breaks, when appropriate to the reconstruction.

Accessibility Considerations

Taking accessibility requirements into consideration early in the planning stage so that barrier-free access can be provided in a way that is compatible with the reconstruction.

Health and Safety Considerations

Considering health and safety code requirements, such as the installation of fire suppression systems, early in the planning stage of the project so that the work is compatible with the reconstruction.

Not Recommended

Installing thermal insulation with a high moisture content.

Using windows and shading devices that are inappropriate to the reconstruction.

Installing new thermal sash with false muntins instead of using sash that is appropriate to the reconstruction.

Removing plant materials and landscape features which perform passive energy functions if they are appropriate to the reconstruction.

Obscuring or damaging the appearance of the reconstructed building in the process of providing barrier-free access.

Meeting health and safety requirements without considering their visual impact on the reconstruction.

Energy, Accessibility, Health & Safety

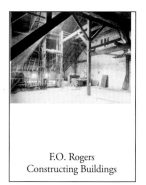

F.O. Rogers
Constructing Buildings

Technical Guidance Publications

The Heritage Preservation Services, National Park Service, conducts a variety of activities to guide federal agencies, states, and the general public in planning and undertaking project work on historic buildings, structures, sites, objects, and districts listed in the National Register of Historic Places. In addition to establishing standards and guidelines, the Division develops, publishes, and distributes technical information on responsible approaches to treating significant historic properties. Guidance materials are offered in several different formats and series, including Program/Training Information, Preservation Briefs, Technical Reports, Preservation Case Studies, and Preservation Tech Notes. The following were published in Washington, D.C., unless otherwise noted.

These books, handbooks, technical leaflets and videos are available through sales from several outlets, including the Superintendent of Documents, Government Printing Office, the National Technical Information Service, and the Historic Preservation Education Foundation. A Catalog of Historic Preservation Publications with stock numbers, prices, and ordering information may be obtained by writing: National Park Service, Heritage Preservation Services, P.O. Box 37127, Washington, D.C. 20013-7127.

Standards and Guidelines

While the Standards provide a consistent philosophical framework for treatment, the Guidelines suggest a model process to follow in the work, and thus assist in applying the Standards to both historic buildings and landscapes.

The Secretary of the Interior's Standards for Rehabilitation (1990). Codified as 36 CFR 67. Leaflet.

The Secretary of the Interior's Standards for the Treatment of Historic Properties (1995). Codified as 36 CFR 68. Leaflet.

The Secretary of the Interior's Standards for Rehabilitation with Guidelines for Rehabilitating Historic Buildings (1990).

The Secretary of the Interior's Standards for Rehabilitation with Illustrated Guidelines for Rehabilitating Historic Buildings (1992).

Working on the Past with The Secretary of the Interior's Standards for the Treatment of Historic Properties: Understanding Preservation, Rehabilitation, Restoration, and Reconstruction (video, VHS). Kay D. Weeks, with Jim Boyd and Mike Quinn (1995).

Program/Training Information

The Heritage Preservation Services prepares a number of publications—both free and sales—that describe or explain program activities (such as the landscape initiative, grants-in-aid, or federal surplus property transfers) or are produced as a result of outreach training, workshops, and study groups.

A Heritage so Rich. Leaflet.

America's Landscape Legacy. Leaflet.

Federal Land-to-Parks Program & Historic Surplus Property Program. Leaflet.

Historic Preservation Fund Grants-in-Aid. Leaflet.

Preserving the Past and Making it Accessible for People with Disabilities. Booklet.

Accessibility and Historic Preservation: Entrances to the Past (video, VHS). Kay D. Weeks and Jim Boyd, with Kay Ellis and David Park. 1993.

Accessibility and Historic Preservation Resource Guide. Thomas C. Jester and Judy Hayward, Compilers. 1992.

Directory of Cultural Resource Education Programs. Compiled and edited by Emogene Bevitt and Heather L. Minor with others. 1995.

Federal Historic Preservation Laws. Sara K. Blumenthal, revised by Emogene A. Bevitt. 1993.

Preserving and Revitalizing Older Communities: Sources of Federal Assistance. Leslie Slavitt. Edited by Susan Escherich. 1993.

Second Lives: Architectural Study Collections in the United States. Emogene A. Bevitt. 1994.

Skills Development Plan for Historical Architects in the National Park Service. Hugh C. Miller, FAIA, Lee H. Nelson, FAIA, and Emogene A. Bevitt. 1988.

A.S Sherwood Lightning Rod

Preservation Tax Incentives

The Heritage Preservation Services develops guidance and certification procedures for the Preservation Tax Incentives Program, provides oversight of regional decisions on historic rehabilitation projects and hears appeals of certification denials.

Historic Preservation Certification Application (includes 36 CFR 67, The Secretary of the Interior's Standards for Rehabilitation).

Preservation Tax Incentives for Rehabilitating Historic Buildings. Leaflet.

Tax Incentives for Rehabilitating Historic Buildings: Current Fiscal Year Analysis.

Affordable Housing Through Historic Preservation: A Case Study Guide to Combining the Tax Credits. Susan Escherich and William Delvac. 1994.

Guidance on Historic Preservation Treatments

The Heritage Preservation Services produces and distributes a variety of technical and educational publications to promote the maintenance and long-term preservation of cultural resources. This 25-year collection includes a wide variety of publications, including NPS Reading Lists (annotated bibliographies), Preservation Briefs, Technical Reports, Preservation Case Studies, Preservation Tech Notes, and Co-Published Books.

NPS Reading Lists

Bibliographies compiled and annotated by preservation experts provide essential background for undertaking responsible work on historic properties.

Historic Building Interiors, Volumes I and II. Compiled by Anne E. Grimmer. 1994.

Preserving Historic Lighthouses. Compiled by Camille M. Martone and Sharon C. Park, AIA. 1989.

Maintaining Historic Buildings. Compiled by Kaye Ellen Simonson. 1990.

Preserving Historic Building Materials: Wood, Paint, Masonry, Concrete, Twentieth Century Materials (1900-1950). Anne E. Grimmer, General Editor. 1993.

Making Educated Decisions: A Landscape Preservation Bibliography. Edited by Charles A. Birnbaum, ASLA, with Cheryl Wagner. 1994.

Pioneers of American Landscape Design: An Annotated Bibliography. Edited by Charles A. Birnbaum, ASLA, with Lisa Crowder. 1993.

Preservation Briefs

Preservation Briefs assist owners and developers of historic buildings in recognizing and resolving common preservation and repair problems prior to work. The briefs are especially useful to preservation tax incentive program applicants because they recommend those methods and approaches for rehabilitating historic buildings that are consistent with their historic character.

Preservation Briefs 1: The Cleaning and Waterproof Coating of Masonry Buildings. Robert C. Mack, AIA. 1975.

Preservation Briefs 2: Repointing Mortar Joints in Historic Brick Buildings. Robert C. Mack, AIA, de Teel Paterson Tiller, and James S. Askins. Rev., 1980.

Preservation Briefs 3: Conserving Energy in Historic Buildings. Baird M. Smith, AIA. 1978.

Preservation Briefs 4: Roofing for Historic Buildings. Sarah M. Sweetser. 1978.

Preservation Briefs 5: The Preservation of Historic Adobe Buildings. 1978.

Preservation Briefs 6: Dangers of Abrasive Cleaning to Historic Buildings. Anne E. Grimmer. 1979.

Preservation Briefs 7: The Preservation of Historic Glazed Architectural Terra-Cotta. de Teel Patterson Tiller. 1979.

Preservation Briefs 8: Aluminum and Vinyl Siding on Historic Buildings: The Appropriateness of Substitute Materials for Resurfacing Historic Wood Frame Buildings. John H. Myers. Revised by Gary Hume. 1984.

Preservation Briefs 9: The Repair of Historic Wooden Windows. John H. Myers. 1981.

Preservation Briefs 10: Exterior Paint Problems on Historic Woodwork. Kay D. Weeks and David W. Look, AIA. 1982.

Preservation Briefs 11: Rehabilitating Historic Storefronts. H. Ward Jandl. 1982.

Preservation Briefs 12: The Preservation of Historic Pigmented Structural Glass (Vitrolite and Carrara Glass). 1984.

Preservation Briefs 13: The Repair and Thermal Upgrading of Historic Steel Windows. Sharon C. Park, AIA. 1984.

Preservation Briefs 14: New Exterior Additions to Historic Buildings: Preservation Concerns. Kay D. Weeks. 1986.

Preservation Briefs 15: Preservation of Historic Concrete: Problems and General Approaches. William B. Coney, AIA. 1987.

Preservation Briefs 16: The Use of Substitute Materials on Historic Building Exteriors. Sharon C. Park, AIA. 1988.

Preservation Briefs 17: Architectural Character: Identifying the Visual Aspects of Historic Buildings as an Aid to Preserving Their Character. Lee H. Nelson, FAIA. 1988.

Preservation Briefs 18: Rehabilitating Interiors in Historic Buildings: Identifying Character-Defining Elements. H. Ward Jandl. 1988.

Preservation Briefs 19: The Repair and Replacement of Historic Wooden Shingle Roofs. Sharon C. Park, AIA. 1989.

Preservation Briefs 20: The Preservation of Historic Barns. Michael J. Auer. 1989.

Preservation Briefs 21: Repairing Historic Flat Plaster—Walls and Ceilings. Marylee MacDonald. 1989.

Preservation Briefs 22: The Preservation and Repair of Historic Stucco. Anne E. Grimmer. 1990.

Preservation Briefs 23: Preserving Historic Ornamental Plaster. David Flaharty. 1990.

Preservation Briefs 24: Heating, Ventilating, and Cooling Historic Buildings: Problems and Recommended Approaches. Sharon C. Park, AIA. 1991.

Preservation Briefs 25: The Preservation of Historic Signs. Michael J. Auer. 1991.

Preservation Briefs 26: The Preservation and Repair of Historic Log Buildings. Bruce D. Bomberger. 1991.

Preservation Briefs 27: The Maintenance and Repair of Architectural Cast Iron. John G. Waite; Historical Overview by Margot Gayle. 1991.

Preservation Briefs 28: Painting Historic Interiors. Sara B. Chase. 1992.

Preservation Briefs 29: The Repair, Replacement, and Maintenance of Slate Roofs. Jeffrey S. Levine. 1992.

Preservation Briefs 30: The Preservation and Repair of Historic Clay Tile Roofs. Anne E. Grimmer and Paul K. Williams. 1992.

Preservation Briefs 31: Mothballing Historic Buildings. Sharon C. Park, AIA. 1993.

Preservation Briefs 32: Making Historic Properties Accessible. Thomas C. Jester and Sharon C. Park, AIA. 1993.

Preservation Briefs 33: The Preservation and Repair of Stained and Leaded Glass. Neal A. Vogel and Rolf Achilles. 1993.

Preservation Briefs 34: Applied Decoration for Historic Interiors: Preserving Historic Composition Ornament. Jonathan Thornton and William Adair, FAAR. 1994.

Preservation Briefs 35: Understanding Old Buildings: The Process of Architectural Investigation. Travis C. McDonald, Jr. 1994.

Preservation Briefs 36: Protecting Cultural Landscapes: Planning, Treatment, and Management of Historic Landscapes. Charles A. Birnbaum, ASLA. 1994.

Preservation Briefs 37: Appropriate Methods for Reducing Lead-Paint Hazards in Historic Housing. Sharon C. Park, AIA, and Douglas C. Hicks. 1995.

Preservation Briefs 38: Removing Graffiti from Historic Masonry. Martin E. Weaver. 1995.

Technical Reports

Technical Reports address in detail problems confronted by architects, engineers, government officials, and other technicians involved in the preservation of historic buildings.

A Glossary of Historic Masonry Deterioration Problems and Preservation Treatments. Anne E. Grimmer. 1984.

Cyclical Maintenance for Historic Buildings. J. Henry Chambers, AIA. 1976.

Epoxies for Wood Repairs in Historic Buildings. Morgan W. Phillips and Dr. Judith E. Selwyn. 1978.

Gaslighting in America: A Guide to Historic Preservation. Denys Peter Myers. 1978. Re-issued by Dover Publications, New York. 1990.

Keeping it Clean: Removing Dirt, Paint, Stains, and Graffiti from Historic Exterior Masonry. Anne E. Grimmer. 1987.

Metals in America's Historic Buildings: Uses and Preservation Treatments. Margot Gayle, David W. Look, AIA, and John G. Waite, AIA. 1980. Revised 1992.

Moisture Problems in Historic Masonry Walls: Diagnosis and Treatment. Baird M. Smith, AIA. 1984.

Moving Historic Buildings. John Obed Curtis. 1975. Reprinted 1991 by W. Patram for the International Association of Structural Movers, Elbridge, New York.

Photogrammetric Recording of Cultural Resources. Perry E. Borchers. 1977.

Rectified Photography and Photo Drawings for Historic Preservation. J. Henry Chambers, AIA. 1973.

Using Photogrammetry to Monitor Materials Deterioration and Structural Problems on Historic Buildings: Dorchester Heights Monument, A Case Study. J. Henry Chambers, AIA. 1985.

X-Ray Examination of Historic Structures. David M. Hart. 1975.

Preservation Case Studies

Preservation Case Studies provide practical solution-oriented information for developers, planners, and owners by presenting and illustrating a specific course of action taken to preserve one building or an entire block of buildings. Individual case studies may highlight an innovative rehabilitation technique, financing strategies, or an overall planning methodology. The majority of the Preservation Case Studies were published in 1978, 1979, and 1980 as exemplary "completion reports" that documented various aspects of Historic Preservation Fund (HPF) grant-in-aid project work.

Abbeville, South Carolina: Rehabilitation Planning and Project Work in the Commercial Town Square. John M. Bryan and the Triad Architectural Associates. 1979.

Fort Johnson, Amsterdam, New York: A Historic Structure Report. Mendel-Mesick-Cohen. 1978.

Main Street Historic District, Van Buren, Arkansas: Storefront Rehabilitation/Restoration Within a Districtwide Plan. Susan Guthrie. 1980.

Maymont Park - The Italian Garden, Richmond, Virginia: Landscape Restoration. Barry W. Starke, ASLA. 1980.

The Morse-Libby Mansion: A Report on Restoration Work, 1973-1977. Morgan W. Phillips. 1979.

Olmsted Park System, Jamaica Pond Boathouse, Jamaica Plain, Massachusetts: Planning for the Preservation of the Boathouse Roof. Richard White. 1979.

Planning for Exterior Work on the First Parish Church, Portland, Maine, Using Photographs as Project Documentation. John C. Hecker, AIA, and Sylvanus W. Doughty. 1979.

Rehabilitating Historic Hotels: Peabody Hotel, Memphis, Tennessee. Floy A. Brown. 1979.

White House Stone Carving: Builders and Restorers. Lee H. Nelson, FAIA. 1992.

Preservation Tech Notes

Preservation Tech Notes provide innovative and traditional solutions to specific problems in preserving cultural resources—buildings, structures, and objects. Tech Notes are intended for practitioners in the preservation field, including architects, contractors, conservators, and maintenance personnel, as well as for property owners and developers.

Exterior Woodwork No. 1: "Proper Painting and Surface Preparation," by Sharon C. Park, AIA. (1986)

Exterior Woodwork No. 2: "Paint Removal from Wood Siding," by Alan O'Bright. (1986)

Exterior Woodwork Number 3: "Log Crown Repair and Selective Replacement Using Epoxy and Fiberglass Reinforcing Bars," by Harrison Goodall. (1989)

Exterior Woodwork No. 4: "Protecting Woodwork Against Decay Using Borate Preservatives," by Ron Sheetz and Charles Fisher. (1993)

Finishes No. 1: "Process-Painting Decals as a Substitute for Hand-Stencilled Ceiling Medallions," by Sharon Park, AIA. (1990)

Doors No. 1: "Historic Garage and Carriage Doors: Rehabilitation Solutions," by Bonnie Halda, AIA. (1989)

Historic Interior Spaces No. 1: "Preserving Historic Corridors in Open Office Plans," by Christina Henry. (1985)

Historic Interior Spaces No. 2: "Preserving Historic Office Building Corridors," by Thomas Keohan. (1989)

Masonry No. 2: "Stabilization and Repair of a Historic Terra Cotta Cornice," by Jeffrey Levine and Donna Harris. (1991)

Masonry No. 3: "Water-Soak Cleaning of Historic Limestone" by Robert Powers (1992).

Mechanical Systems No. 1: "Replicating Historic Elevator Enclosures," by Marilyn Kaplan, AIA. (1989)

Metals No. 1: "Conserving Outdoor Bronze Sculpture," by Dennis Montagna. (1989)

Metals No. 2: "Restoring Metal Roof Cornices," by Richard Pieper. (1990)

Metals No. 3: "In-kind Replacement of Historic Stamped-Metal Exterior Siding," by Rebecca Shiffer. (1991)

Museum Collection Storage No. 1: "Museum Collection Storage in a Historic Building Using a Prefabricated Structure," by Don Cumberland, Jr. (1985)

Museum Collections No. 1: "Reducing Visible and Ultraviolet Light Damage to Interior Wood Finishes," by Ron Sheetz and Charles Fisher. (1990)

Site No. 1: "Restoring Vine Coverage to Historic Buildings," by Karen Day. (1991)

Temporary Protection No. 1: "Temporary Protection of Historic Stairways," by Charles Fisher. (1985)

Temporary Protection No. 2: "Specifying Temporary Protection of Historic Interiors During Construction and Repair," by Dale Frens. (1993)

The Window Handbook: Successful Strategies for Rehabilitating Windows in Historic Buildings. Charles E. Fisher, Editor. Tech Notes included in the Window Handbook are as follows:

Planning and evaluation:

Windows No. 1: "Planning Approaches to Window Preservation," by Charles Fisher. (1984)

Windows No. 10: "Temporary Window Vents in Unoccupied Historic Buildings," by Charles Fisher and Thomas Vitanza. (1985)

Repair and Weatherization:

Windows No. 14: "Reinforcing Deteriorated Wooden Windows," by Paul Stumes, P. Eng. (1986)

Windows No. 16: "Repairing and Upgrading Multi-Light Wooden Mill Windows," by Christopher Closs. (1986)

Windows No. 17: "Repair and Retrofitting Industrial Steel Windows" by Robert Powers. (1989)

Double Glazing Historic Windows:

Windows No. 2: "Installing Insulating Glass in Existing Steel Windows," by Charles Fisher. (1984)

Windows No. 5: "Interior Metal Storm Windows," by Laura Muckenfuss and Charles Fisher. (1984)

Windows No. 3: "Exterior Storm Windows: Casement Design Wooden Storm Sash," by Wayne Trissler and Charles Fisher. (1984)

Windows No. 8: "Thermal Retrofit of Historic Wooden Sash Using Interior Piggyback Storm Panels," by Sharon Park, AIA. (1984)

Windows No. 9: "Interior Storm Windows: Magnetic Seal, by Charles Fisher. (1984)

Windows No. 11: "Installing Insulating Glass in Existing Wooden Sash Incorporating the Historic Glass," by Charles Fisher. (1985)

Windows No. 15: "Interior Storms for Steel Casement Windows," by Charles Fisher and Christina Henry. (1986)

Replacement Frames and Sash:

Windows No. 4: "Replacement Wooden Frames and Sash," by William C. Feist. (1984)

Windows No. 6: "Replacement Wooden Sash and Frames With Insulating Glass and Integral Muntins," by Charles Parrott. (1984)

Windows No. 12: "Aluminum Replacements for Steel Industrial Sash," by Charles Fisher. (1986)

Windows No. 13: "Aluminum Replacement Windows with Sealed Insulating Glass and Trapezoidal Muntin Grids," by Charles Parrott. (1985)

Windows No. 18: "Aluminum Replacement Windows with True Divided-Lights, Interior Piggyback Storm Panels, and Exposed Historic Wooden Frames," by Charles Parrott. (1991)

Screens, Awnings, and Other Accessories:

Windows No. 7: "Window Awnings," by Laura Muckenfuss and Charles Fisher. (1984)

Co-Published Books

The National Park Service works cooperatively with outside preservation organizations to produce conference handbooks and other guidance materials on appropriate treatments for historic properties.

Interiors Handbook for Historic Buildings. Michael J. Auer, Charles E. Fisher, and Anne Grimmer, Editors. National Park Service and Historic Preservation Education Foundation. 1988.

Interiors Handbook for Historic Buildings, Volume II. Michael J. Auer, Charles E. Fisher, Thomas C. Jester, and Marilyn E. Kaplan, Editors. National Park Service and Historic Preservation Education Foundation. 1993.

Preserving the Recent Past. Deborah Slaton and Rebecca Shiffer, Editors. National Park Service and Historic Preservation Education Foundation. 1995.

Respectful Rehabilitation: Answers to Your Questions on Historic Buildings. Kay D. Weeks and Diane Maddex, Editors. National Park Service and National Trust for Historic Preservation. 1982.

Twentieth-Century Building Materials: History and Conservation. Thomas C. Jester, Editor. The National Park Service and McGraw-Hill. 1995.

The Window Directory for Historic Buildings. Charles E. Fisher and Thomas C. Jester, Editors. National Park Service and Center for Architectural Conservation, Georgia Institute of Technology. 1992.

The Window Workbook for Historic Buildings. Charles E. Fisher, Editor. National Park Service and Historic Preservation Education Foundation. 1986.

T.M. Hill
Corn Planter